The
SELF-SUFFICIENCY
HANDBOOK

The
SELF-SUFFICIENCY
HANDBOOK

A COMPLETE GUIDE TO GREENER LIVING

Alan & Gill Bridgewater

Skyhorse Publishing

www.skyhorsepublishing.com

10 9 8 7 6 5 4 3 2 1

The Library of Congress Cataloging in Publication Data

Bridgewater, Alan.
The self-sufficiency handbook : a complete guide to greener living / by Alan & Gill Bridgewater.
p. cm.
Includes index.
ISBN-13: 978-1-60239-163-5 (alk. paper)
ISBN-10: 1-60239-163-7 (alk. paper)
1. Environmental protection—Citizen participation. 2. Environmentalism. 3. Organic living.
I. Bridgewater, Gill. II. Title.

TD171.7.B745 2007
640—dc22
2007024094

Editorial Direction: Rosemary Wilkinson

Senior Editor: Anne Konopelski

Production: Hazel Kirkman

Design: Glyn Bridgewater

Illustrations: Gill Bridgewater

Editor: Alison Copland

Photographs: AG&G Books

Printed in China

CONTENTS

INTRODUCTION

When Gill and I met at art school in the 1960s, the whole place was buzzing with a new kind of freedom. Somehow, we all felt that we could do it – meaning life – better than previous generations. I remember one evening sitting in a college common room listening to two young, hippy, American lecturers animatedly talking about how very soon we would all be forced by the failure of oil supplies to return to some sort of Amish type self-sufficiency – log cabins maybe, horses rather than cars, communes where groups of like-minded people pulled together to create a better society – and it was very exciting. As they saw it, and as whole swathes of people saw it, our consumer society was living off the fast-shrinking capital resources of the earth. Their thinking was that ever since the start of the industrial revolution we had been taking and dumping: taking the coal and dumping the waste, taking the oil and dumping pollution, taking the goodness from the soil and leaving it barren, cutting down trees, and so on.

The big question at that time was 'How can we change from being a greedy, grabbing, take-it-and-run, dirty, despoiling society to a greener, more giving society?' The general consensus of opinion in art schools and universities was that the best way forward was not to try to change things from the center, but rather, as Timothy Leary said in the 1960s, to 'turn on, tune in, drop out,' meaning look inward at yourself, look outward at society, and then, in the light of your knowledge, select and reject. The idea was that changes were best made from the edges in.

My reading list at that time was topped by two books: *Cottage Economy* by William Cobbett and *Walden* by Henry Thoreau. *Cottage Economy*, first published in 1821, is one of the very first self-help manuals, in that it describes in blow-by-blow detail how 'a large part of the food of even a large family may be raised … from forty rods, or quarter of an acre.' In *Walden*, published in 1854, Thoreau describes how, on deciding that he was going to set up his self-sufficient home in a hut in the woods, he gave thought to the implications of every minute detail, from the orientation of the hut in relation to the sun through to how he could survive on fish and beans. Further to this, however, Thoreau attacked the way things were going at that time – the rip-tear-slash-and-burn farming, the development of the railroads and the growth of the cities, all at the expense of nature. Gill and I read these books avidly; we found it so stimulating – the notion of being completely independent with no public electricity, no public water, no public anything!

By the time the 1960s were out, we were married and living in a tumbledown house in the middle of a field with no running water, no electricity, no mortgage … in fact, nothing much at all except our two toddler sons, and a clear and certain knowledge that we were going to be self-sufficient. It all seemed so beautifully simple: I would continue teaching pottery, Gill would do her weaving, and along the way we would sort out the house, establish a craft workshop, dig a well, build a windmill, grow our own food, have chickens, and generally live happily ever after. As I saw it at that time, our progress from bare plot through to being self-sufficient was something like being Robinson Crusoe on his desert island. The general idea was not that we would drop out and go back to some sort of pre-industrial, rural, horse-pulling basics, but rather that we would take from the best that was on offer – like Crusoe did from his foundered ship – and use it to create a new world. All a bit romantic, I know, but that was the way we felt.

So it was that, when we were living at Valley Farm, we tried to look at every problem analytically through Thoreau-Cobbett-Crusoe eyes. We looked long and hard at the problem, and then over the following days and weeks we did our best to figure out how to put it right. For example, when it came to sorting out the water, we looked at the existing well, we pumped it dry, we timed how long it took to refill, we had the water tested, and then, in the light of the fact that the water was failing and grossly polluted, we went through various stages of collecting and storing rainwater, fitting various pumps and filters, having a borehole drilled, and so on until the problem was solved. It was the same with the waste water, the livestock and everything else – we looked at the problem, we did the research, we talked to old folk who had experienced living off-grid, we considered how each change or procedure would impact on the environment, and then shaped our life accordingly. Along the way, of course, we could not help but be influenced by our past experiences. For example, when I was a boy, living with my grandparents in the country, I remember thinking that most people lived happy, self-sufficient lives without the need for running water, cars, TVs and washing machines.

Later, in the 1970s, we were influenced by two more books, *Self-Sufficiency* by John Seymour and *The Autonomous House* by Robert and Brenda Vale, as well as by *The Last Whole Earth Catalogue* and *Harrowsmith*. *The Autonomous House* is also a wonderful book – dynamic, forward-thinking and altogether inspirational.

Naturally, we had many failures along the way – stock dying, running out of money, taking advice from the wrong people, and such like, but every experience – good and bad – has added to the sum total of our knowledge. When we first started out, people like us were damned with the label 'hairy hippies,' but gradually, with the 1970s oil crisis, and the various American back-to-earth movements, and 'The Good Life' TV program, the media and the establishment gradually came to accept that there was another way. As to what precisely that way was

or is, all most people knew for sure was that we were burning up irreplaceable resources, and poisoning ourselves in the act.

Here we are 30–40 years later, and have things significantly improved? The oil supplies are running out, there are more cars than ever, there is more pollution, the forests are being hacked down at a faster rate, people are stressed out by fast-paced living, and mass-produced, low-quality food is making people ill. On the positive side, however, whereas back in the 1960s terms like 'ecological,' 'eco-friendly' and 'recycling' labelled the user as being some sort of weird, vegetarian egghead, they are now high up in people's thinking.

Really practical questions are being asked. How can we go off-grid? How can we heat our homes without gas and oil? How can we grow food without using chemicals? How can we maximize our recycling? The good news is that not only do we have the answers, but governments and think-tank groups are also urging people to go self-sufficient. For example, when I was looking around for a wind turbine back in the early 1970s there were only one or two very expensive, hit-and-miss machines on the market; now there are dozens, perhaps hundreds, of wind turbines to choose from. Better yet, there are government grants!

Just in case you are wondering, self-sufficiency is not about becoming all long-faced and wordy about the theoretical possibilities – not a bit of it. It is about rolling up your sleeves, and having fun, cutting living costs, eating better, and generally doing your bit for the 'green revolution.' Of course, we know about not having enough cash, and not having enough space, but – to our way of thinking – if everyone made some small change then we would be three parts there. If ever there was a time for self-sufficiency, this is it.

We hope that you will find this book both helpful and inspirational as you start out along the exciting path to self-sufficiency.

THE LAND

THE PERFECT PLOT

If you wish to become completely self-sufficient, it is no good just casually playing around with the idea; you are going to have to become passionate about the whole notion – obsessional even – to the point where it begins to be a fantasy that takes over your whole life. When this creative and inspirational moment is reached, you can start working out how you are going to turn your imaginings into reality. When you and your loved ones are well and truly committed to the idea, then comes the wonderfully exciting, even mind-bending, business of thinking about what is possible.

QUESTIONS TO ASK YOURSELF

- How much land do you need?
- Is your dream plot in the city or the country? Or even another country where the land and property are less expensive?
- Do you want to try to go off-grid – meaning no public electricity, water or gas?
- Can you make it happen by staying put and renting land – fields, allotments and such?
- Do you want to join a self-sufficient community?
- Are you going to continue a career?
- Do you have enough assets?

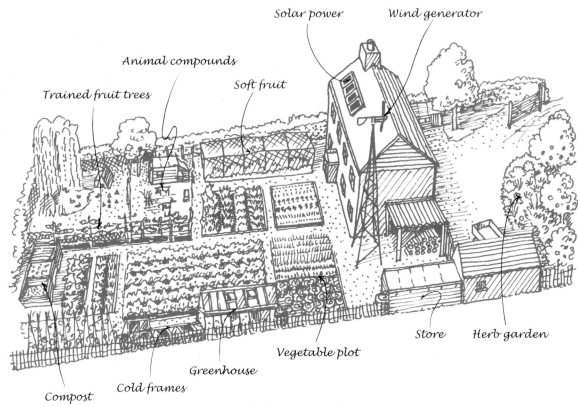

A dream plot

Solar power · Wind generator · Animal compounds · Soft fruit · Trained fruit trees · Store · Herb garden · Vegetable plot · Greenhouse · Cold frames · Compost

HOW MUCH LAND DO YOU NEED?

The amount you will need will relate to its location, the climate, the quality of the land and your needs. Maybe a single person could just about manage with half an acre of rich land – if they were going to grow all their own food, or grow cash crops and trade and swap for other essentials – but to my way of thinking two people in this day and age would need more like two acres. Then again, if they opted for coppiced woodland and sold say poles, posts or turned items as a cash product, perhaps it would work out. If your particular dream plot involves fishing, horses, sheep, chickens, allotments or whatever it might be, you will already have a pretty good idea of how much land you need.

For example, our first plot was just over an acre of remote land with poor, boggy ground and difficult access. We were in our late twenties with two babies. It succeeded in a sort of messy fashion, simply because I earned money as a lecturer. The regular income made everything possible. Gill was able to stay at home and do her weaving, I made pottery in the long summer holidays, and along the way we both looked after our two boys, and kept a few chickens, geese, sheep and goats. Of course it was not easy, but it worked for us because we lived off-grid and were not weighed down with utility bills.

TRIAL PERIOD

So you want to live in the woods, or by the sea, or … and you are a bit nervous about burning your boats. The good news is that you do not have to – it is much better to have a trial period. Say that you own or are buying a house in the city, and you have a dream that involves moving to Spain and growing almonds. The prudent way forward is to rent your city property, and to use the cash to rent a place in your chosen area. In this way you will get to know the area and the people. You won't be living the complete dream, but you will be learning about the location. You will be able to have a close-up look at the countryside. You will be able to experience the local weather patterns, and talk to farmers and hear what they have to say about the land and the local markets, and learn about the schools, and so on. If at the end of the trial period you discover that the area is not quite right, you can move back and start again.

CAN A BACKYARD BE A PERFECT PLOT?

The answer is most definitely yes. If you take a look at the illustration opposite you will see how even a small yard can be put to productive use. Look at your plot and start by drawing up a detailed what-where-and-how year-round plan. You will have to be well organized, and you will have to make sure that every corner is put to good use, and you will have to be super-selective when it comes to livestock. There won't be room for a cow, but you might be able to have a goat, half a dozen hens, and perhaps even a beehive or two. Your location might also be ideal for selling produce such as organic eggs, honey and goat's cheese to your neighbors at the gate.

THE BEST TIME TO BUY

The very best time to buy a property is in late summer – when the owners are beginning to be concerned that they might not sell before winter. If you can get in with a low offer, when the owners are getting anxious about winter costs, then there is a very good chance that you will get a bargain. But don't be too clever, because if you leave it too late and wait until, say, the last few weeks running up to Christmas, then the chances are that the owners will have made plans for the holiday season, and will have more or less decided to wait for the spring and the promise of higher prices. It is always a good idea to have your money arrangements set up and at the ready, so that when the time is right you can dive straight in and make a swift deal.

ORIENTATION

In Britain, Europe and America, the ideal is for the house and land to be set so that it is backed on the north side by rising ground – so that the front of the house and the best ground is on a gentle south-facing slope. I can think of two small neighboring set-ups both next to a ridge of land – a high road that runs from east to west. The two houses are set so that they look across the road at each other. They are both very successful enterprises, but the house on the south side of the road, the one that has the house and land on the south side of the ridge, is better placed because the

back of the house, and all the land and gardens, are sheltered from the north winds and looking to the sun. The house on the north side of the ridge, on the other hand, is at a disadvantage because the winds roll up the hill and hit the back of the house.

Interestingly, when you really look at the two properties in the 'flesh,' and at old maps, you can see that these lie-of-the-land advantages and disadvantages are borne out by house names, and by farming activities. For example, the place on the north side of the ridge is called 'High Winds,' while the place on the south side is called 'Sunny Side.' The place on the south side has had orchards for the last hundred or so years, while the farm on the north side has always had lush meadows. They are both good options; it is just that one place has it slightly easier than the other.

When the time comes that you are searching for a property in earnest, look at the maps, walk over and around the land, talk to neighbors, look at the way a place is operated, look at the buildings, look at the

Check the orientation of the plot

way the trees are growing – for example are they windswept or covered in moss? – and generally find out as much as you can about the property you are considering.

THE BEST SIZE?

Of course much will depend upon your particular dream, and the size of your bank balance, but the best size has always got to be … as much as possible! The truth is that, when you really get down to it, most green-dreamers are usually channeled by circumstances into going for a rural property that has land, rather than say a small farm. For the most part, they will opt for a house with say 3–5 acres. Experience tells me that most properties of this type tend to be marginal land.

What this all adds up to is that, while you have to start out with a dream, you also have to be prepared to make compromises. The thing to do is look long and hard at the not-quite-right options and see if you can work around the obvious problems. The house might be in a mess – but that can be fixed. Or there is simply too much land – but that can be rented out. One place we saw was far too expensive, but when we did our research we found out that at least two houses backing onto the property were keen to extend their gardens by buying bits of the land, so much so that the original high price could easily be offset by selling little pieces of the land.

Above all, it is vital to be open-minded. A certain place might not be big enough, but perhaps you could rent additional land. Don't become disenchanted at this point; when we were looking for our first place, our newspaper ad ('Wanted to buy – isolated property with no near neighbors') resulted in over 20 replies, one of which came up trumps.

OWN WATER SUPPLY

You must have water. You can use public water, take water from a well or borehole, build a cistern to save rainwater, or take water from a river or stream that runs across your land.

PUBLIC WATER This is of course an easy if expensive option, especially if you have livestock. If you have no choice, then the best you can do is use it carefully, and see if you cut costs by backing it up with one of the other off-grid options.

WELL WATER The good news is that the very existence of a well suggests that it was once used. Pump the well dry, and time how long it takes to refill. This will give you some idea of quantity. Make contact with a local health department and have the water tested.

BOREHOLES It is not only expensive to have a borehole drilled, but worse still you cannot be sure at the end of it that you will strike water, or that the water will be fit to drink. Some borehole companies are prepared to work on a pay-if-find basis.

CISTERNS A cistern is no more or less than a tank for catching and storing water. A cistern is a good option if you live in an area of heavy rainfall. A good middle way forward if your place has an existing system is to use it to cut the costs of using public water.

STREAMS AND RIVERS You can in many instances take water from a river or stream that runs across your land, the proviso being that you must ask the appropriate authorities and have the water tested (the latter is particularly vital).

FRIENDS AND FAMILY OPTIONS

One way forward if you are a bit low on capital, or you have in mind to go for a big set-up that needs a larger workforce, is go in with friends or family – a bit like your own mini commune. Of course, much depends on your unique situation and circumstances, but let us say for the moment that you are an average couple and you are short on money, and one or other set of parents would like to stump up part of the funds and throw their lot in with you. This is a very common scenario. It is not right for everyone, but there are lots of advantages. Older people have more life experience, older people have more money, and older people generally have more free time, and so on. In addition, lots of tasks are much easier if there are more than two pairs of hands.

Going in with friends is much more complicated. Experience tells me that it is always best to start such an arrangement by building in an escape route – so that dissatisfied members of the group can pull out. There have to be written ground rules, and the whole arrangement must be put on a legal footing.

This property, consisting of a main house and an annex, is large enough for two families to join forces in self-sufficiency

THOUGHTS AND QUESTIONS

- You might well need your parents' or partner's parents' help with the money, but can you work together? Have you ever lived and worked together?
- If you do have plans to go in with family or friends, you would all have to make compromises. You would each need to have your own areas of responsibility, but at the same time there would need to be a single captain.
- It is important that there is plenty of living space so that all members can pull back into their own private areas.
- If you go in with parents, what happens when one dies? Will the property need to be sold? How will such an arrangement affect other members of the family?
- If you go in with friends, what happens if one party wants to pull their money out?
- A good option is to buy a big place so that say two couples can live in their own houses. Is this possible?

RIGHTS OF WAY

The ideal is to have clear-cut legal access to your property and land with ownership set out in your deeds. The next best option is to have a legal right to get from the road through to your property. As for footpaths, it is best if your land is free of them.

My advice – the only advice – is to make sure right from the start that you have legal rights that are shown, plotted and registered on your deeds. As to public footpaths, make sure that the arrangement of the land is such that they won't hold you back if say you want to keep livestock. If you have any doubts about rights of way, visit the local records office and ask to see the appropriate maps.

NO NEAR NEIGHBORS

Neighbors in the city can be a problem – too close and too many, and sometimes an item that you just don't need or want. Neighbors in the country, on the other hand, are a must. Of course, you don't want to be living in a neighbor's pocket, but, sure enough, the time will come when you need their help, or they need yours.

A difficult scenario is when your house and land used to be part of a much larger set-up, to the extent that maybe your little cottage is sitting next to a large farmhouse, your land is encircled, and more experienced folk are always on hand to give advice. This is most definitely a bad arrangement. You will need help and advice, but it is important that you are allowed to a great extent to make your own mistakes. You don't want to have someone always looking over your shoulder, and you don't want to put yourself in a position where you might feel intimidated.

OFF-GRID LAND

Although the self-sufficiency dream is all about going off-grid – no public electricity, gas or water – this ideal cannot easily be achieved right from the start. For example, with our first place, we had the house and barn complete with a vast rainwater tank, all on 1 acre, and that was about it. We knew there was a well, but it was hidden away somewhere in the undergrowth.

We more or less camped in the barn for the first year – drinking and washing using the rainwater, cooking and lighting with bottled gas, and heating with an open fire. It can be done, but is a lot of hard work. A much better scenario is to start out with public services, and then gradually replace each of them with 'eco' options.

COUNTRY LIVING

Mud, mud and more mud!

If you were born on a farm, you will know what country living is all about, but it is not so easy to fully appreciate the good, bad and ugly possibilities if you have always lived in a city or small town. Let us say, then, that the scenario is that you are a town person, and you dream about being self-sufficient in the country. You have drawn inspiration from Thoreau's 'Walden,' you have thought it all through, you have done your sums, and you are really beginning to make plans. You want to leave the gray city life behind, get back to what you consider is a simple way of life, and generally to live the dream. I will take it that you have driven round the countryside searching out various options, and I would hope that you have had a trial dry run living in the country.

The not-so-strange thing is that many people are so wrapped up in romantic notions of living in the country that they forget about the sometimes harsh realities. To my way of thinking, country living is by far the easiest go-green option – land is less expensive, there is more space, and there is more choice – but there are many town dwellers who are self-

sufficient. Country living in the UK would also be very different to country living in the backwoods of rural Australia, for example. It is vital that you sum up all the advantages and disadvantages before you make the move. Be mindful that some of the advantages and disadvantages are different sides of the same coin.

DISADVANTAGES

- Living in the country can be very lonely in the first instance, especially if you are seen as being different.
- Country people can be very standoffish and wary if they perceive you as being sophisticated or uppity.
- Transport costs can be high, especially if you live off the beaten track. You must work this into your budget.
- You will need a good reliable van or truck, possibly a four-wheel drive. A breakdown in the country can be very difficult if you are non-mechanical – dangerous even.
- Stock (pigs, goats and so on) need 24-hour care. You will still have to feed and tend the animals, even on celebration days.
- Illness! A cold, or flu, or a sprained ankle can be a huge problem, especially if you have livestock.
- School-age children have to get to school – this can mean a long round-trip twice a day, or educating the children at home, or even sending them to boarding school.
- Weather in the country will make more of an impact – snow, rain, drought and wind can and will rule your life.
- Country living and darkness go hand in hand. A country lane or a place out in the wilderness will be completely dark at night.

- Mud! Mud will get everywhere.
- Living in a small community – a hamlet or a small isolated village – can be very restricting. Everyone around will know your business.
- Town pets – dogs and cats – sometimes find it impossible to settle down to living in the country. If your dogs chase sheep or are allowed to roam, you will fall out with your neighbors.
- Fences and gates are a constant cause of friction. If you leave a gate open, stock will run loose, and you will get into trouble. The simple rule with a gate is close it after you if it you find it closed, and leave it tied-back open if that is the way you found it.

Feeding pigs on celebration days

ADVANTAGES

- The feeling of space can be spiritually uplifting for many people.
- You will almost certainly derive pleasure from the changing seasons and vistas.
- Generally, there is less pollution in the country. Some farmers still use chemical sprays, but at least you won't have to battle with car fumes.
- If you give the locals a chance, they can be very caring and protective, especially if they see you as a struggling agricultural worker who is simply trying to do his or her best.
- Noise pollution is low – much lower than in the city. You will be able to hear yourself think. You will be able to hear birds singing, your animals calling, the wind in the trees, and so on.
- Light pollution is low. You will be able to see the stars at night.
- If you have school-age children, you will soon be drawn into school activities. Your kids will bring friends home, you will meet their parents, and so on. You will soon know most kids and parents in your area.
- Land costs, meaning agricultural land, are low. You can rent whole fields, spreads or woods.
- Rural activities can be good fun and low in cost. For example, if your kids want to build a camp in the fields, great – they cannot do this in the town.
- You will have room to do your thing: take up riding, watch nature, build eccentric garden structures, run around the garden singing … whatever turns you on.
- If you are of an independent nature, you will feel empowered.
- There are fewer people. You will have more personal space. This can be very important for some people.

TOWN LIVING

The simple question you have to ask yourself is this: is it possible to go self-sufficient in the town? The answer is yes. It is very different to trying to make a go of it in the country, but it is a very real option. I think it fair to say that, while lots of town and country folk dream about going green in the country, it is only town folk who think about going green in the town. As for whether or not the town option is a second best choice – because they cannot make a go of it in the country so they will have to see if it works in the town – the facts and figures suggest that for many people the town scenario is their first choice.

Then, of course, there are towns and towns. A small town in middle England is going to be very different from a small town in Canada or Australia. For example, I can think of a town in Australia where the population is so low and the ribbon development so sprawling that there are bits of spare land just about everywhere. I can also think of a large town in Canada where there are areas of farmland right in the middle of the community. That said, no matter the country, the overall differences between town and countryside are going to be broadly the same.

ALLOTMENTS

Allotments are a very good option in the UK. They are not only extremely low in cost, but better yet there are still plenty on offer. We know of a couple who, with five plots between them, are completely self-sufficient in vegetables and fresh eggs – and anything they cannot eat they sell. Much depends upon the area, but most allotment associations traditionally allow the holders to keep small livestock – chickens, rabbits, goats and such – as long as they are securely fenced and housed.

LARGE GARDENS

A good-sized garden is a good option, especially if you turn the total garden over to food production. You will not be able to have a polytunnel, unless it is a really huge garden – because the neighbors will be sure to object – but there is nothing to say that you cannot have one or more greenhouses. We have friends who make it all work by breeding rabbits in the large walled backyard, growing food on an allotment, and selling surplus in the front yard. The rabbits are grown and reared in large sheds.

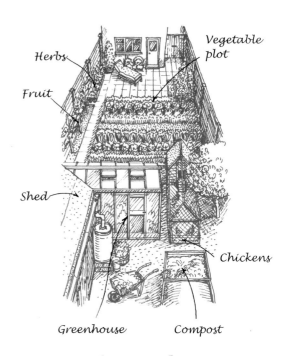

Herbs

Vegetable plot

Fruit

Shed

Chickens

Greenhouse

Compost

Town garden

RENTED GROUND

You can always rent small pieces of ground – the neighbors' gardens, small islands of land that have somehow been forgotten, land at the edge of town, and so on.

DISADVANTAGES

- Living in the town can be very lonely.
- Town people generally will not much care about what you are at, as long as it does not impinge upon their space.
- Transport costs can be high – especially if you need to travel out to the country for essential supplies.
- Just as in the country, your stock – pigs, goats and all the rest – need 24-hour care. You will still have to feed and tend the animals, even on celebration days.
- The problem with keeping stock in the town is that things like smells and noise will be an issue.
- To an ever-growing extent, mindless vandalism and town living go hand in hand. This could be a problem if you have a flourishing allotment, or you keep chickens, for example.
- A self-sufficient set-up in the town has to be a tighter and more controlled operation than one in the country, simply because space is at a premium.
- There is more pollution in the town. Land is more expensive.
- There will be restrictions – how many chickens, how much noise, and so on.

ADVANTAGES

- The frenetic activity – the feeling that you are at the heart of it – can be inspirational.
- You will be able to draw inspiration from cultural activities – museums, art galleries, theaters, lectures, and so on.

Selling your produce

- The ground is less likely to be affected by severe frosts.
- You can get by with a beaten-up car or van, simply because there will always be garages and mechanics close at hand.
- You will be able to do part-time paid work to support your go-green activities.
- If you already live in the town, you will not be bothered about relocation costs.
- Allotments are so low in cost that the money will hardly figure. Some allotments even have special rates if you are trying to make a full-time go of it.
- Much depends upon your definition of self-sufficiency, but it is possible to dramatically cut food costs by buying time-dated food that you cannot produce yourself.
- If your notion of self-sufficiency has to do with selling produce or other items to raise cash, then you can sell at markets, garage sales and such like. The large population gives you a large market. So, for example, you will have no problems selling items like goat's milk, cheese and fresh veggies.
- Much depends upon the town, but there are a growing number of community farms – set-ups that use brownfield sites. You will be able to join like-minded groups.
- Your children will have no problems getting to school.

LIVING ABROAD

More and more people are moving abroad – from the UK and USA to places like Spain, France and Goa, for the simple reason that land and property are cheaper and the weather is perceived as being easier. Of course, it is not all sunshine and siestas, but for many people it is a good solid option.

PLANNING AN ESCAPE ROUTE

The best advice is never to move abroad without building in an escape route. So you are young and ready for anything, and maybe you think my advice is a bit over-cautious and parental – I agree, it is – but for all that I still say it is vital, if your venture fails, that you can move back home. Just be aware that today's dream place in the sun can be tomorrow's nightmare. Of course, you cannot and should not plan for every grim what-if, but you should try at the very least to maintain a foothold in your mother country – such as an apartment, a piece of land, a cache of money in the bank. You must manage the move in carefully considered step-by-step stages, making sure that one stage is working out before going on to the next.

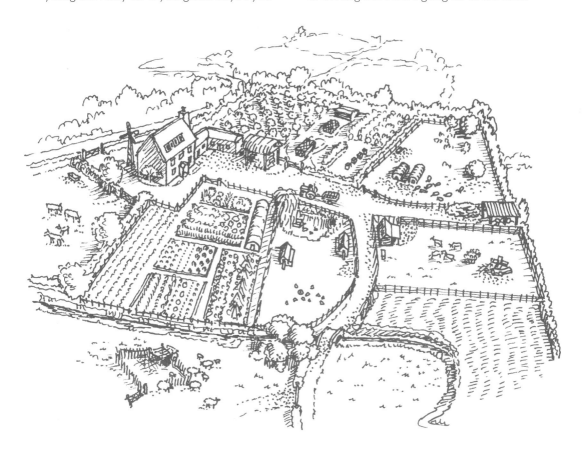

Imagine ... a huge plot of land abroad

I Dream about your move abroad.

2 Have a year-long trial run in the country of your choice.

3 Plan your escape route, so that you can go home if your venture goes wrong.

4 Employ an experienced legal adviser to guide you every step of the way. Make sure you know about your legal rights in the country of your choice.

5 Set up your mother-country home base and rent it out.

6 Move to the country of your choice.

DISADVANTAGES

- Certainly the weather might be milder, but the summers can also be too hot, and it can rain for half the year.
- There are earthquake zones in New Zealand, Greece, India, some parts of Italy, and so on. You must check this out.
- The biggest problem for most people moving abroad is that they cannot speak more than a few words of the language.
- Different countries have different cultures. This is not so bad if, say, you are English-speaking and move to England, Canada, Australia or even a country that has long associations with England and America, such as Italy, Greece or Portugal, but moving to a place like Vietnam, Goa or South America is very different.
- Laws and rights can be very different from one country to another. Question – can you legally buy land in a place like Goa? Be warned that you cannot always take it for granted that you can sell a property in, say, Central America and take all the proceeds from the sale out of the country.
- Health care can be a problem in some parts of the world. You must check this out if you have ongoing special health needs.

- Building standards can range from good to terrible. Your chosen haven might be a mess of bad plumbing, non-existent drains, on-and-off electricity/water supplies, poor phone lines, and so on.
- Moving abroad can be a culture shock.
- Schools can be very different, even impossible, in some parts of the world. Your children's formal education will almost certainly be set back in the short term, but their overall experience will be enriched, and they will soon have a second language. Have you thought about things like exams and college, however?
- Some countries have very low standards of health care and poor hospitals.
- The land and property might be cheaper, but the living standards might be lower.
- Much depends upon your chosen country, but there might be local resistance to you – a foreigner – buying land.

ADVANTAGES

- The food can be wonderful – lots of fresh fruit and veggies, cheap local wines, locally caught fish, and so on.
- Some parts of the world are blessed with a wonderful climate.
- We know of a couple who live in Portugal and they report that the weather is mild. They also say that the people are more relaxed – or at least their perception is that the people are more relaxed.
- In some parts of rural France property is very low in cost. That said, property laws are very complex.
- If you want to move within the European Union – say from England to France – and you want to take over a vineyard, or grow food, or whatever, then there are resettlement grants.

21

THE SELF-SUFFICIENT HOUSE

THE PERFECT HOUSE – AN OVERVIEW

Growing your own food

The self-sufficient house is just that – a house that is self-contained and independent, a house that generates its own power and recycles its own waste. But there is more to it than that. In the perfect self-sufficient scheme of things, the people living in the house would not be going out to work to earn money and then spend the money on food, but rather they would stay at home and spend all their time growing their own food. Of course, most people now agree that this self-sufficient dream of completely dropping out – like some sort of New-Age Robinson Crusoe – cannot happen.

The reality is that we all need money to pay taxes, housing rates and all the rest. The perfect self-sufficient house can only be a part of a much larger picture. Ironically, whereas in the 1960s and 1970s most governments tended to regard go-greeners as slightly eccentric, they are now pushing us all to do our bit … to use less energy, produce less waste, drive smaller cars, use less water, insulate our houses, and so on.

The truly wonderful thing is that many of the much-debated visionary possibilities of the 60s and 70s are now realities. You don't have to dream about such items as small wind generators, low-energy cars, solar heating, etc.; you can now buy them off the shelf. The truth is that if we all do our bit to go self-sufficient – some doing no more than saving energy by insulating their homes, others trying to go the whole hog and work towards going off-grid – then the country, the world even, will go a long way to solving its ever-pressing energy needs. Better still, where most people once had no choice other than to travel to work, now the invention of the computer and all its linked technologies has meant that more and more people can now opt for doing a good part of their work at home.

Most self-sufficiency gurus are now of the opinion that the best option is to take the middle way. Their thinking is that most of us do need to earn some money, but the bulk of our time could be spent growing our own food and servicing our self-sufficient houses. All that said, the reality is that most people will

Rooftop wind energy system

not by choice go down the self-sufficiency road, they will not be interested in any part of the go-green package, but they will by necessity be forced to use 'eco' technology to cut living costs.

THE MAKE-DO-AND-MEND HOUSE

The make-do-and-mend solution is aimed at four groups of people: those who by default choose eco-green technology simply because they need to cut costs and they are not interested in anything other than that; those who buy into green technology because it is fashionable; those who are really keen to go green but have to stay put and make the best of what they have got; and those who are inspired to move house and do their best to be self-sufficient, but are still forced to go for a little-by-little approach. The make-do-and-mend philosophy involves a step-by-step DIY approach and making the best of a bad job. The way forward with this option is to look at what you have – whether a large or a small house – and then make a series of changes to cut energy costs.

MAKE-DO-AND-MEND STEP-BY-STEP WORK PLAN

1　Fit good heavy curtains to all the windows and exterior doors. Open the curtains during the day and close them at night to keep the heat in. Wear layered clothes when the weather is cold.

2　Replace all single-pane windows with double- or even triple-glazed units.

3　Reduce the number of exterior doors.

4　Insulate the roof space and all cavity walls. If you have solid walls, add insulation to the inside or outside faces.

5　In the northern hemisphere, reduce the number or size of north-facing windows and increase the size and number of south-facing windows; vice versa if you live in the southern hemisphere.

6　Install a woodburning stove, and remove items like gas and electric fires. Fit vents and ducts so that you can channel excess heat around the house.

7　Fit a glass house, glass porch or conservatory over every exterior door. Fit vents at floor and ceiling height, so that you direct hot air from the glass structures into the house, and so that you can create a cooling circulating system.

8　Fit solar collectors on your roof, so you can preheat your cooking and heating water.

9　Modify your water system to save and reuse 'gray water' (see page 42), so you can use your bath and shower water to flush the toilets.

Make-do-and-mend house

THE PERFECT HOUSE – THE AUTONOMOUS HOUSE

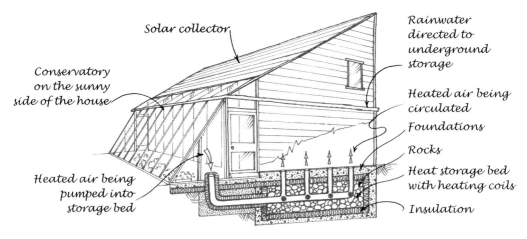

Solar collector

Rainwater directed to underground storage

Conservatory on the sunny side of the house

Heated air being circulated

Foundations

Rocks

Heated air being pumped into storage bed

Heat storage bed with heating coils

Insulation

Autonomous house (Brenda and Robert Vale)

When Brenda and Robert Vale wrote their ground-breaking book *The Autonomous House* back in the 1970s, they inspired and encouraged a whole generation to go off-grid. I remember one evening sitting in a candlelit college common room – the oil crisis was in full swing and petrol had more than doubled in price – listening to a group of art students excitedly talking about how we could all become self-sufficient. The bit in the Vales' book that really got me going was where an autonomous house was likened to a sort of land-based space pod which was designed to provide an environment that was free from the existing life-support structures of earth. The impact of the book was such that the term 'autonomous house' is now commonly used to describe a particular type of practical and proactive house and environment set-up. *The Autonomous House* is still in print 30 or so years after it was first published – a perfect recommendation if ever there was one.

THE ANATOMY OF AN AUTONOMOUS HOUSE

An autonomous house is best thought of as a total environment where every single energy-creating and waste-recycling aspect is considered and modified at a very basic intuitive level; more than that, however, it is an environment where the relentless aim is to sever all links with utility pipelines and power cables. The autonomous house has solar collectors on the roof, systems to move hot air around the house, rock beds to store heat, water-storage systems, wind generators to turn the power of the wind into electricity, simple systems to recycle waste, small-scale gas plants, systems to save 'gray water', systems to save store and use rainwater, systems that use fuel cells, and so on. The autonomous approach requires that you should proactively challenge the status quo, and tirelessly battle away making changes, until you achieve a stand-alone off-grid environment.

THE PERFECT HOUSE – THE PASSIVE HOUSE

A passive house is a house that uses no designated energy systems for the central heating, but rather has passive systems that gather waste heat from the domestic hot-water system, from cooking and lighting, and from the bodies of the people living in the house, plus heat from the sun. In this approach, the building itself or elements of or within the building are designed so that they take advantage of natural solar heating.

Operable windows, vents, Trombe walls, thermal chimneys and insulation are the primary elements found in passive design. Operable windows are simply windows that can be opened, while Trombe walls use materials such as masonry and water that can store heat energy. The sun shines through the glazing and heats the masonry walls or tanks of water, so that the space between the glazing and the wall becomes a thermal chimney. Vents set at floor and ceiling level in both the glass and the interior Trombe walls are managed so that the currents of hot air that rise by convection between the wall and the glass – in the thermal chimney – are directed either in or out of the building. Depending on the time of year, and the choice of open and closed vents, the rising hot air can be used either to heat or to cool the space.

Key aspects of passive design include appropriate solar orientation, the use of thermal mass, very high levels of insulation, and air locks or thermal buffers on exterior doors. The thermal mass absorbs heat during the day and gives it out at night. Ideally, a passive house is a long, thin structure, with total glazing on one of the long sides, with the house sited so that the glazed side is facing the sun at midday. Passive systems are simple, have few moving parts, and require minimal maintenance.

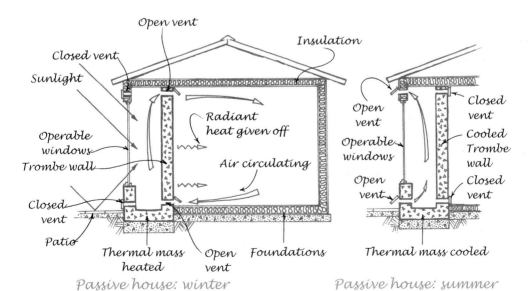

Passive house: winter

Passive house: summer

THE PERFECT HOUSE – THE HIGH-TECH HOUSE

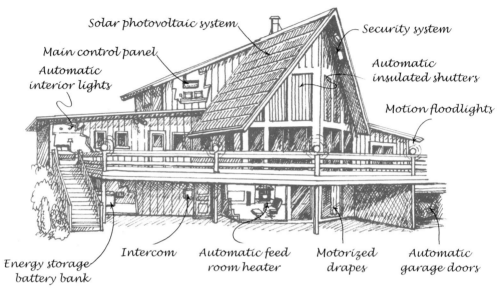

Solar photovoltaic system

Security system

Main control panel

Automatic interior lights

Automatic insulated shutters

Motion floodlights

Energy storage battery bank

Intercom

Automatic feed room heater

Motorized drapes

Automatic garage doors

A fully automated high-tech house

A high-tech house, sometimes called a 'smart house' or even an 'automated house,' is a house that contains automated computerized devices that control the systems within the house. There are devices that will open and close doors, turn lights on and off, open and close blinds and shutters, open and close vents and valves, and so on. Of course, the high-tech house is unsuitable if you are a technophobe, and the set-up costs are high, and certainly the whole idea has been heavily criticized as being somehow or other not quite green (and in many ways this is true) but, once it is up and running, the high-tech approach is a very good way of cutting energy costs to the bone.

Overall, the high-tech approach involves reducing energy costs by trimming away at every aspect of waste, and by fine-tuning

usage. Maybe a system that turns valves and lights on and off does not sound so wonderful, but if a batch of systems, or better still a single completely integrated system, were to, as it were, follow you around the house opening and shutting down the lights, temperature, ventilation and the like – then you can see that there are great cost-cutting possibilities.

The high-tech or smart system is designed to respond to outside stimuli. For example, if you link it up with, say, a passive solar heating system, temperatures outside the house, and your day-to-day time spent in the house, then it is capable of fine-tuning the various controls to a remarkable degree. Apart from cutting costs, and making all manner of everyday tasks easier, such a system is a great option if you are aged or in any way physically handicapped.

THE PERFECT HOUSE –
THE NATURAL HOUSE

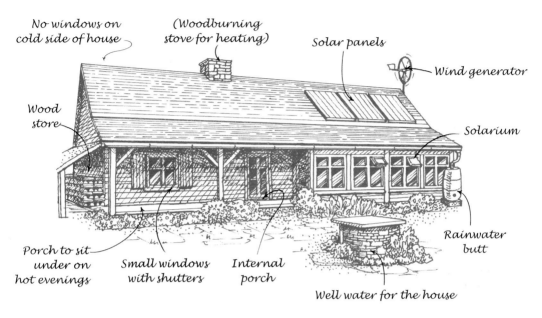

No windows on cold side of house

(Woodburning stove for heating)

Solar panels

Wind generator

Wood store

Solarium

Porch to sit under on hot evenings

Small windows with shutters

Internal porch

Well water for the house

Rainwater butt

A natural house uses natural building materials

A natural house is one where every aspect of the house – its form, structure, materials and setting – are completely in harmony with the environment. The whole notion of a natural house consists of trying to get back to 'the way it was.' If you look at traditional folk and tribal houses from all around the world, you will see that, by using local materials and technologies that have been tried and trusted for hundreds of years, and by placing buildings so that they have their backs to the wind and their faces to the sun, the people were more in tune with nature. In the winter whole families sat around a central fire, while in the summer they sat outside under covered porches. Rooms were small and dark, but the buildings did at least work. The scale was

right, and the color and texture related to the environment; the house was healthy and in tune with nature.

Once the Industrial Revolution was up and running, the various communities around the world rejected their tried and trusted building methods in favor of new and easier options. Thatched roofs were replaced by slates, tiles and tin sheet, thick mud walls were replaced by brick, asbestos and corrugated iron, houses were set down in neat and narrow rows without a thought to the direction of the sun or to prevailing winds, and so on.

The natural house involves looking afresh at old, sound, traditional principles. The key words here are small, natural, local materials, passive, healthy, spiritual and harmony.

HEATING AND COOKING – TRADITIONAL OPTIONS

ELECTRICITY FROM THE GRID

Most of us in the developed countries are on grid electricity. Most of us have electric lights in every room, perhaps an electric cooker, microwave ovens, toasters, kettles, a television, small electric heaters, lots of electric power tools, computers, maybe a pump to run an oil system, and so on ad infinitum – all powered by electricity.

FOR
- Installation costs are relatively low, and once the system is in place – all cables and sockets – it will last for 25 years or more.
- Electricity is instant, clean, silent and very convenient – you just flip a switch and it is up and running.
- Electricity is relatively safe – it is not going to leak out or be stolen.
- Public electricity is always there on demand – you don't have to remember to order it.
- The prices of on-grid electricity are government regulated – you don't have to worry about costs spiralling out of control.
- Governments will always try to ensure that electricity needs are met.

AGAINST
- You have to be linked to the grid – this might be difficult or even impossible in some isolated areas.
- You have to pay a standing charge, even though you might make efforts to reduce your consumption.
- Electricity is the most expensive fuel.

GAS FROM THE GRID

In times past, piped or on-grid gas was considered to be a miracle fuel; first we had coal gas and then gas from the various oilfields. It was once used for lighting, but now it is used primarily for heating and cooking. Though just a few years ago on-grid gas was considered to be a low-cost option – a good option for central heating – reduced supplies are rapidly forcing costs up.

FOR
- Though initial installation costs are relatively high, once the system is in place – the boiler and all the pipes – it will more or less last indefinitely.
- On-grid gas is instant and very convenient – you just turn on the taps.
- Gas is relatively is safe – as long as it is fitted by trained specialists.
- Gas is always there on demand – you don't have to remember to order it.

AGAINST
- You have to be linked to the grid – this is not viable in some isolated areas.
- Currently gas prices are rising rapidly, to the extent that it is fast becoming a prohibitively expensive option.
- You have to pay a standing charge no matter what your consumption is.
- Gas can be used for lighting, but the lighting level produced is below recommended standards.
- Gas is potentially very dangerous – it is both toxic and explosive.

OIL

First it was oil lamps, and then oil for combustion engines, and then oil for domestic central heating systems, and then oil for everything! Over the last hundred years our consumption has gone mad – low-cost air travel, and of course billions of cars … we have been throwing it away. So here we are in the first years of the second millennium AD, and oil is fast running out. Oil is great for heating and cooking, but how long will it last?

FOR
- Though the initial set-up costs are high, once the system is in place it will last for a long time.
- Oil is very convenient.
- Some cookers, stoves and boilers are very stylish – for example, the Aga range cooker – they look and feel good.

AGAINST
- You are at the mercy of the delivery truck – if it cannot reach you, you are in trouble.
- Supplies worldwide are dwindling and costs spiralling.
- There was an oil shortage in the 1970s and there is a threat of another one on the way. It is unsettling.
- You need to have a huge tank in the garden in which to store the oil.

COAL

Coal was king back in the nineteenth and twentieth centuries – every house had a couple of open coal fires, and perhaps even a coal cooking range. When I was a kid in the 1950s, family life was centered around the open fire: lighting it in the morning, filling up the coal scuttle, making tea and toast, polishing the brass. Every house puffed out black smoke. But that all came to an end with the various Clean Air Acts. Now, with oil and gas costs rising, the time is fast coming around when coal is once again being thought of as a viable and practical energy source.

FOR
- In the UK at least there are vast coal reserves – enough to last at least 400 years.
- All the dangers associated with burning coal are understood – we know all about CO emissions.
- There are no intangibles as there are with nuclear fuel.
- It is possible to turn coal into easy-to-handle fuels.
- It is now possible to burn coal cleanly.

AGAINST
- Coal is bulky – transport and storage are a problem.
- Coal mining is both dirty and dangerous.
- Using coal for heating and cooking would involve a huge political turn-around on the part of governments.
- Coal can be burnt cleanly on a domestic scale, but the set-up would be costly.
- Using coal in the house – even if it is in a 'smokeless' form – is both dusty and dirty.

A coal-fired range

WOOD

Wood is a good traditional option both for space heating and cooking – wonderful if you live in the country and are attracted to all the back-to-the-land associations, but what if you live in the suburbs? It is a viable option for some part of the population because they live near a forest, but it is not an option for the majority, if for no other reason than that it would not be possible to turn that amount of land over to sustainable forestry.

FOR

- Woodburning stoves and boilers are intrinsically attractive – they appeal to our back-to-the-land traditions.
- It does involve a lot of physical effort, but some people are attracted by the hands-on, back-to-basics activity.
- In some areas, woodburning is a very good option – it is available, low in cost and sustainable, and the technology is understood.
- You can opt for a stove that has an open door and a hotplate if you enjoy the additional pleasures of making tea and toast.

A woodburning cooker

AGAINST

- Even at its best – if you live in the country and wood is readily available – it is still messy and dusty.
- You must have plenty of storage space – lots of sheds and covered areas.
- Wood involves a lot of physical effort – moving it from the shed, cutting it to a usable size, taking it into the house, lighting the fire, removing the ash, and so on.
- Wood is anything but instant – its use involves a lot of forward planning.

FUEL AND ENERGY RECAP

In the 1970s, when the oil crisis was in full swing, and we were creeping around with candles, heating oil was difficult to find, and schools, industry and shops were being closed for some part of the day, there was a huge amount of interest in alternative energy. Books and articles were written, governments set up all sorts of committees, energy grants were given, and individuals went off and did their own thing. There was a feeling of urgency. The general consensus was that something had to be done fast before the non-renewables ran out. All sorts of doomsday figures and timetables were given and mooted. Some said that that the gas and oil would run out in 50 years.

The only thing that all parties agreed on was that something had to be done; and that was where it became difficult. Although it was agreed that we had to search out sustainable and renewable energy sources, and we had to do it fast, governments were loath to take responsibility. What actually happened in the end was that new oil reserves were found, oil prices fell, and the whole situation was forgotten, or at least pushed to the back of our minds.

The strange thing is that, while there is no denying that the oil and gas are running out, consumption is going up. The situation is getting even more complicated, too, with the rich and powerful countries now rushing to get their hands on the remaining oil and gas reserves. That is not the worst of it, either. Not only have we squandered our energy, but in doing so we are also damaging our environment – some say irreversibly.

MAKING SMALL CHANGES

Our reliance on non-renewable fossil fuel – coal, gas and oil – is mad, bad and dangerous. We know that it has got to stop. We know that we cannot go on spending our energy capital. That said, it is also plain to see that we cannot easily go back to the way it was. Most people are well and truly hooked into using their car, and buying food out of season, and buying cheap clothing from abroad, and all the other activities that are eating away at our global energy reserves. So what can we do?

The answer is for each of us to try as individuals to cut back on our use of fossil fuels in favor of renewable natural energy sources – sun, wind, earth and water. We cannot make the change in one fell swoop, but we can at least try to achieve a balance by generating as much energy as we can from the sun, wind, water and earth. And it is possible. Let us say, for example, that at this moment you are running your home on gas from the grid – a gas-fired boiler for heating and cooking. You could start by super-insulating the inside walls and attic of your home. If the recommended thickness in the attic space is 6 in., you could increase it to 12 in. You could fit thicker curtains. You could install a log-fired stove for heating the space, and you could fit solar panels on the roof.

These measures might seem a bit tame, but the end result would be that the greater part of your energy needs would come from renewable sources. Of course, if we all made lots of little changes, then we would be three parts there.

USING LESS ENERGY IN AND AROUND THE HOME

- Wear suitable clothes according to the season. For example, if you are cold put on another layer of clothes rather than turning up the heating.
- Wear more energy-efficient natural fibers – wool to keep you warm and cotton and linen to keep you cool.
- Improve the insulation in the loft, in the walls and under the floor.
- Fit double or triple glazing to windows.
- Hang thick insulated curtains in the winter.
- Fit draught-proofing around all the doors and windows.
- In the winter, improve solar gain by having the curtains drawn well back in the daytime and tightly closed at night.
- In the summer, open windows/draw blinds to keep cool – rather than turning on electrical cooling systems.
- Cut back on your use of electrical appliances – food mixers, power tools, hair driers, electric blankets, and all the other gadgets that use energy.
- Settle for lower lighting levels. Only light the room or even the small space you are using.
- Shelter the house on the windward side with trees, hedges, wooden panels.
- Fit porches or conservatories on the exterior doors.
- Walk to the local shops, school, church or friends' houses.

WOODBURNING STOVES

There is nothing quite so comforting on a wet and windy winter's night than to be sitting in front of a glowing woodburning stove. A good, modern stove will warm a large living area – the proviso being that it is correctly fitted and matched to the room – and will burn for twice as long, and twice as clean, on the same amount of wood, as one of those old cast-iron horrors.

Traditional pot-belly stove

WOODBURNING TIPS

- Hardwoods like oak and maple burn better than softwood. It is a good idea to split the logs so that they dry out more quickly.
- Only burn well-seasoned wood, which contains just 20–30 percent water, because it burns more efficiently than freshly cut wood. Well-seasoned hardwoods provide a long-lasting, high-temperature fire.
- Never burn plastic waste (such as packaging) on the stove, because it will create potentially dangerous fumes.
- Make sure that the chimney/flue size is at least as big as the stove's outlet.
- Wood smoke is potentially harmful, so fit a modern stove that is capable of

decreasing the level of harmful emissions by up to 90 percent.
- Efficient burning is indicated when the smoke looks white or steamy, as opposed to gray or black.
- Be wary about 'damping down' for the night. It does enable you to keep the fire in, but it is one of the factors that create a build-up of creosote in the chimney.
- It is always a good idea to open all the air vents at start-up, so that the resultant brisk fire burns away creosote and pre-warms the chimney.
- Always refer to the manufacturer's guidelines, since design and air-flow systems vary from one stove to another.

LIGHTING A WOODBURNING STOVE

1 Crumple up half a dozen sheets of crisp-dry newspaper (ordinary cheap newspaper, not glossy magazine paper).

2 Take about 20 pieces of dry split kindling (preferably a softwood like pine) and build a raft-like layer on top of the newspaper.

3 Take two small crisp-dry half-logs (logs that have been split down the middle) and set them down flat on the raft of kindling.

4 With the door(s) wide open, and the air controls open, light the newspaper at as many points as possible.

5 Leave the doors and vents open until the logs are well alight.

6 Allow the fire to burn vigorously and brightly until the logs are well charred, then add more logs and cut the draught down in stages, or according to the manufacturer's instructions.

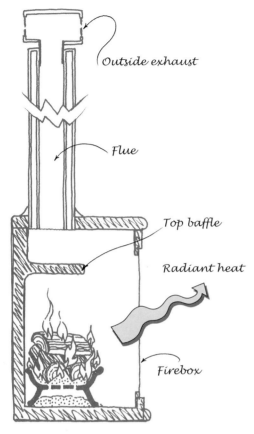

Outside exhaust

Flue

Top baffle

Radiant heat

Firebox

Cross-section of a stove

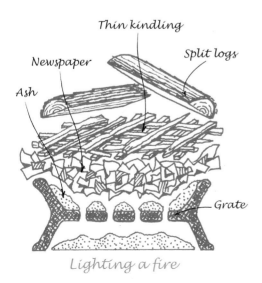

Thin kindling

Newspaper

Split logs

Ash

Grate

Lighting a fire

GLOSSARY OF WOODBURNING STOVES

BAFFLE Wall around which the smoke and gas must flow in its passage through the stove; wall designed to control the air flow.

CREOSOTE Sticky mix of high flammable tars and oils that can build up within the smoke pipe and chimney lining and then ignite, causing a chimney fire – the result of overly damping down.

DAMPING DOWN Act of closing down the vents so as to keep the fire in overnight. Damping down hastens the build-up of potentially dangerous creosote.

DOWNDRAUGHT System of baffles and vents that results in the gases being forced down through the body of the fire before they are allowed to go up the chimney.

EMISSIONS Harmful by-products. A good modern stove (see left) produces relatively low emissions.

FIREBOX The containment that holds the body of the fire.

FLUE The chimney, pipe, or vent.

HEARTH The fireproof base (usually made of concrete, tiles or metal) on which the stove stands.

KINDLING Thin dry wood (usually softwood) that is used to start a fire.

SEASONED Term used to describe logs that have been allowed to dry for 6–12 months. Seasoned wood burns more efficiently than freshly cut wood.

SECONDARY AIR System that introduces a stream of air above the body of the fire, which results in the gases being forced down through the fire before being allowed up the flue.

LOOKING AT YOUR FLUE SYSTEM

Once you have decided that you want to fit a woodburning stove, and assuming that you have an existing chimney and fireplace, have a good look at the way they are arranged. You will be faced with two options. You have to choose a stove to fit what you have; or, if you like a stove that does not fit, you have to change the chimney to suit the stove. Either way, if you are living in an old house with a small open fire, or a new house that has a chimney but has never been fitted with a stove, the chances are that the flue or chimney will be a simple brick structure. If this is the case, it will need to be lined.

WHY LINE AND INSULATE YOUR CHIMNEY?

Flue gases from burning wood produces tar and impurities that will in time condensate on the inside surface of the flue. Because a stove is more efficient than an open fire, there will be a greater build-up of tars. Tar build-up is a problem on three counts: it will gradually leach through and stain your walls; it will eventually run down the chimney as a sticky residue and maybe ooze onto the stove; and it will increase the risk of a chimney fire.

Fitting a chimney with an insulated twin-walled liner sorts out all these problems at a stroke; not only does it create a smooth surface that makes it difficult for tars to form, but, if they do form, they are directed back into the fire to be recombusted. Better yet, the consistent diameter of the liner and the fact that it is insulated result in the gases passing through the system at a greater speed – or, to put it another way, the fire will have a better draw and be altogether more efficient.

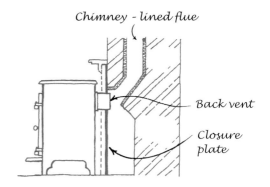

Chimney - lined flue

Back vent

Closure plate

Back-vented stove

CHOOSING A BACK-VENTED STOVE TO FIT A CONVENTIONAL OPEN FIREPLACE

If you have a small conventional open fireplace with a hearth (meaning that you have to kneel down in order to look up the chimney), then one option is to fit a back-venting stove, with a short length of flue that extends horizontally from the back plate directly into the fireplace opening.

1 Measure your fireplace opening – its width, and the height from the surface of the hearth to the top of the opening. Also measure the width/depth of the hearth.

2 Buy a back-vented stove that can be fitted so that it sits flush against the wall with the flue spigot running straight back into the fireplace opening.

3 If your fireplace opening is very low, you might have to run the flue back through the wall at a higher level. If this is the case, choose a stove that allows you to make a new opening at some point higher than the existing fireplace lintel/arch.

4 Once you have chosen your stove – with a flue at a level that fits the existing opening, or you have to cut a new flue hole at some point above the existing opening – sit it in place and clamp it to the flue liner.

5 Finally, seal the fire opening with a metal closure plate or with brickwork. Either brick up the opening or, if you are fitting a metal plate, fit the plate over the pipe before clamping it to the back flue.

FITTING A TOP-VENTED STOVE IN A HIGH-LEVEL COTTAGE-STYLE FIREPLACE

If you have a large high-level inglenook-type open fireplace (you can more or less stand in it and look up the chimney), then you have the choice of either bricking up the opening and fitting a back-venting stove flush to the new brickwork, as just described, so that the stove stands in the room, or you can fit a top-venting stove within the opening, and have a short length of flue running vertically from the stove's top plate up through a metal closure plate and on up the chimney. The object of the closure plate is to seal the opening, so that soot does not fall down, you cannot see the sky, and the heat stays in the room. While the arrangement is good in that it maximizes room space, it is more complicated to fit. If your chimney is really large, you will also need to fill the cavity between the outside of the flexible flue pipe and the inside walls of the chimney with insulation, such as vermiculite. The following step-by-step procedure will allow you to understand how the boiler is to be fitted, either by you or by a specialist.

1 Measure your fireplace opening – its width, the height from the surface of the hearth to the top of the opening, its total depth, and any details such as beams, bulging brickwork and bits of strange ironwork.

2 Visit a stove showroom and choose a top-vented stove that can be fitted into the fireplace opening, either partially in or right up against the rear wall. Go for a design

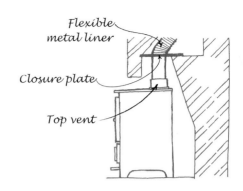

Top-vented stove

that allows the vent pipe to run straight up from the top of the stove and on up the chimney, and a size that allows you to use and service the stove with a good amount of all-round clearance between the top of the stove and the underside of the fireplace opening – if, for example, the stove has some sort of top opening or kettle plate, you need to make sure that it is easy to get to. In a large house, explore the possibility of fitting a stove with a back boiler, so that you can have a hot water or central heating system.

3 Order a metal closure plate to fit the chimney opening. If your fireplace opening is large – perhaps an inglenook-type opening that is big enough to stand in – then the likelihood is that you will need a closure plate that has additional structural members, bracketed supports, access hatches for chimney sweeping, and so on. If this is the case, the best advice is to have your set-up measured by a specialist.

4 Once you have chosen your stove with its custom-made closure plate, sit it in place and clamp it to the flue liner.

5 Finally, seal the fire opening with the metal closure plate, as recommended by the stove manufacturer.

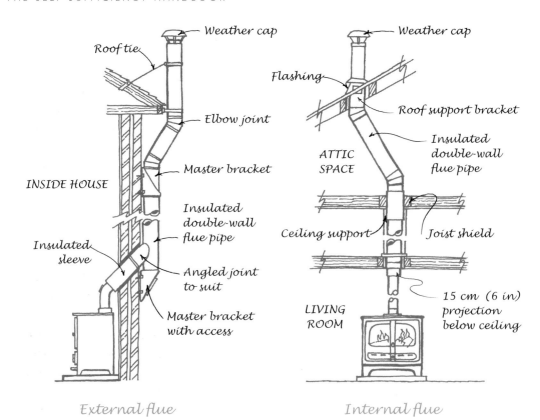

External flue

Internal flue

BUILDING A NEW INSULATED FLUE PIPE CHIMNEY

The scenario is that you want to have a woodburning stove but you don't have a chimney. In this situation the easiest method is to build a chimney using stainless steel double-skin insulated flue pipe. In essence, this is just a system of 5 in., 6 in., 7 in. or 8 in. diameter cylindrical sections – straights, elbows, corners, junctions, brackets, straps, ties – that allow you to build a tubular chimney up to roof level, meaning a chimney either inside or outside the building. The double-skin system results in a Class 1 insulated chimney – a good safe system that conforms to all building codes and regulations. The top-quality powder insulation within the double skin means not only that, to a great extent, the flue gases rise without condensing, but also that you can run the flue up between floors (through tinder-dry loft spaces or through bedrooms) without it being a danger.

What all this adds up to is that, if you want a chimney without working with bricks and mortar, then you can have one. The system needs to be supported at regular intervals, as set out by the manufacturer, by appropriate floor, wall, ceiling and roof brackets and ties. I think that overall the internal installation looks more attractive than the outside option – and it is certainly easier to fit – but that is just a personal view.

WARNING Although the chimney is a relatively basic DIY task, the proviso is that you must use a top-quality double-skin system, follow the guidelines set out by the manufacturer, and follow all the building codes. Some suppliers run a teach-in service.

WOODHEAT SOLAR COLLECTOR

A tree is, at the same time, a perfect solar collector, an ecological miracle and a renewable energy resource. A tree spends its whole life gathering in the sun's energy, and then releasing it through combustion. In the context of burning wood on your stove, you might think that wood is wood is wood – but not a bit of it. Some wood types burn better than others. Put another way, the best wood produces more heat, for a greater length of time, with less smoke and fuss, than inferior woods. So, when you come across an old adage that describes a certain firewood as something like 'sulky,' 'moody,' 'lazy' or 'all smoke and no heat,' then take it to heart. Some wood types are easier to handle and split than others, however.

Firewood can be obtained in two ways: you can order it from a recommended supplier or you can cut it yourself. Either way you need to know what makes a good fire. The following list gives all the relevant details.

ALL YOU NEED TO KNOW ABOUT FIREWOOD

ALDER poor choice – sulky burn – produces little heat – light in weight – little heating value

ASH good choice – produces lots of heat with little smoke – long burn – heavy – medium to high heating value – difficult to saw and split

ASPEN poor to medium choice – medium burn – can be difficult to saw and split

BASSWOOD medium to poor choice – lots of smoke – medium burn – light in weight – easy to split

BEECH medium choice – good amount of heat – medium burn – medium weight – easy to saw and split

BIRCH good choice when dry – produces lots of heat – medium burn – medium weight – easy to saw and split

CEDAR good choice when seasoned – produces fair amount of heat – good smell – makes lots of noise – relatively easy to saw and split

CHERRY medium to good choice – good amount of heat – slow burn – good smell – relatively easy to saw and split

ELM poor choice – not much heat – difficult to burn – difficult to saw and split

HEMLOCK poor choice – not much heat – medium/easy to burn – easy to saw but not so easy to split

MAPLE good choice – long burn and good heat – easy to burn – easy to saw and split

OAK top choice – produces lots of heat with little or no smoke – long burn – heavy – very high heating value – easy to split

PINE poor choice, worse still when wet – short burn with medium heat – easy to burn – lots of noise – easy to saw and split

POPLAR medium to poor choice – medium burn with medium heat – easy to saw and split

WALNUT top choice – burns well with good heat – smells beautiful – difficult to saw and easy to split

WILLOW poor to medium choice – burns better when dry – short burn – produces lots of smoke and not much heat – easy to saw and split

39

LIGHTING

Although most of us take electric light for granted, there was a time, not so long ago, when lighting was noisy, smelly, required lots of effort and was anything but instant.

For example, when my grandpa was a child in the 1890s, they had gas lights, oil lamps, candles and that was about it. Interestingly, however, he told me that a lot of time was spent trimming wicks, fitting new mantles, fetching oil, cleaning the globes and chimneys, buying in more candles and generally keeping the lights in good order, but he did not remember his world as being dark or gloomy. The home, the street, the corner shop, the pub and the like were each sitting in their own individual pools of light, rather than everything being linked by all-encompassing illumination.

It is plain to see that if we reduced our lighting levels we really would not be any worse off. There are alternative ways of making electricity, and these are dealt with on pages 46–61, but just for the moment let us consider how we could make small changes to what we have.

LIGHTS OUT

Since most of us have two or more electric lights in every room, and all manner of other lights in and around the home, it follows that we could save energy simply by only lighting our own personal space. The whole house does not need to be floodlit.

LOW-ENERGY BULBS

We could change all our old bulbs to modern low-energy options. According to the latest figures, this would make a significant nationwide saving.

LIGHTING LEVELS

At the beginning of the twentieth century, various bodies in the UK, Europe and the USA set 'good code' lighting levels for public buildings. For example, in 1910 they thought that about 30 lux – meaning one lumen per square meter times 30 – was a good lighting standard. In 1920 they raised the standard to around 200 lux, then in 1940 to 300+, and so on, until now we have it at about 1,500+. The illogical thing is that anything above about 350 lux means that we are subjected to too much light. Our forebears were straining their eyes trying to read by flickering candlelight, and we are now straining our eyes trying to read under the glare of millions of light bulbs. Some people even have to wear shades indoors, simply because the lighting levels are too high. If we were to settle for a much lower lighting level, but still stay with the recommended level of about 350 lux, we could cut our electric lighting energy consumption by two thirds.

TURN OFF STANDBY LIGHTS

Most modern homes are floodlit at night by a twinkling of little lights telling us that various bits of electrical equipment are ready and waiting to be switched on – the TV, radio, computers, printers, monitors, phones and so on. If you turn them all off until they are needed, another saving could be made.

DECORATIVE FIXES

You could decorate your home with white and cream paint, light-colored fabrics, glossy surfaces, lots of mirrors – anything that makes the best use of available light.

LIVING BY NATURE'S CLOCK

In times past, we more or less got up at dawn and went to bed at sundown. If we got up very early or went to bed very late, we needed some sort of light to illuminate our way. Much the same now goes for various rural communities in third-world countries. For example, if you ask a person from a village in northern India what they do for lighting when

Windows and light-reflecting surfaces

the sun goes down, they will tell you they are so tired that they opt for going to bed. I am not saying that we should all work ourselves to exhaustion so that we go to bed early and save money on lighting, but rather that we could reshape our lives so that there was a better balance between work, sleep and play.

BACK TO THE WAY IT WAS

In the early 1970s we – Gill and I and our two boys – lived for ten years in an isolated farmhouse that was completely off-grid.

I cheated slightly by charging up batteries at the college where I taught – so we were able to have a small amount of TV and radio – but that apart we did not have on-grid electricity. Our lighting was a mix of candles, oil lamps and gas lights. There is a lot of pleasure to be had from sitting around on a summer's evening talking, gathering around a log fire in winter listening to the radio, or simply going to bed early and reading a book by candlelight. It was relaxing and good fun. Of course, every now and then we went out, maybe to the pub or the cinema, but of all the things we missed over those ten years electric lighting was not one of them. I am not suggesting that you cut the power; I am saying that you could change your behavior patterns so as to minimize your lighting needs.

LIGHT TUBES

Light tubes are cylinders, mirrored on the internal face, that are used to reflect sunlight. They are designed to be fitted in the roof. The sunlight strikes the inside face of the top end of the cylinder, with the effect that the light is reflected and bounced down into the building. A light tube will of course only work when the sun is shining, or at least in daylight, but it does mean that you can be working in a loft, a basement or a windowless room without having to have the lights on all the time.

Light tube (cross-section)

WATER

At the beginning of the twentieth century, most people had a well in the garden or, if they were very lucky, a hand-operated pump in the kitchen, scullery or outhouse. Lifting buckets from the well, walking back and forth to the local village pump, pumping water up from the sink to a bucket – the whole business of fetching and carrying water was tedious hard work.

Human nature being what it is, most people did their level best to manage with as little as possible. Figures suggest that most people at that time each used about 1–2 buckets of water per day for everything – cooking, washing, cleaning clothes, and so on. It does not sound like much, but remember that toilet arrangements consisted of not much more than a small outside privy – just a bucket and a jug. Gradually, over the years, with the introductions of the inside bathroom, flushing toilets, showers, washing machines and dishwashers, we have upped our consumption of water from about two buckets a day to somewhere between 30 and 60.

WATER USAGE

If we take it that on average we each use well over 30 buckets of water per day – and this can only be a rough average – figures tell us that about one third of this water, say 10 buckets, are used to flush the toilet. As for saving water, we could save dribs and drabs by cutting back on the washing machine, the dishwasher, washing and hosing the garden, but more than anything else it is plain to see that if we cut back on flushing the toilet the savings would be significant.

GRAY WATER

Gray water is all the domestic water from the bath, shower, kitchen and laundry – in fact just about any water that is free from feces, urine or decomposing food matter. Gray water makes up about 50 percent of the total water that we put down the drain. There is no denying that we need to flush the toilet, but we don't need to do it with top-quality drinking water. If we used gray water to flush our toilets, we would save water and energy on at least two counts – we would save 50 percent of our water costs, and 50 percent of total consumption. The problem with gray water is that it is mostly polluted to some

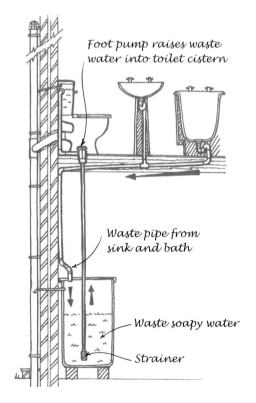

Foot pump raises waste
water into toilet cistern

Waste pipe from
sink and bath

Waste soapy water

Strainer

Gray water installations

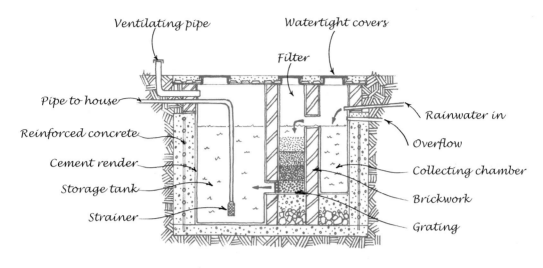

*Domestic sanitation
(devised by Godwin and Downing)*

degree with soaps and detergents. This would not be a problem if it is used to flush the toilet, but it is not so good for putting on the garden. The other difficulty is that gray water soon goes smelly.

One good low-cost option is a little bathroom-sink arrangement that fits on top of a low-level toilet system that allows hand-washing water to be used to flush. You use the toilet, flush, and then wash your hands. The next person uses the toilet, uses your washing water to flush, and so on. It is a very clever low-tech solution.

WELLS AND BOREHOLES
One of the swiftest ways of ensuring a good supply of clean water, and cutting long-term costs, is to have your own borehole. Water boreholes are environmentally friendly in as much as they cut back on the overall take-out from rivers and reservoirs. In most instances – and this depends upon location – borehole water is completely independent from surface

water. A hole is drilled to a depth of 50+ yards and lined with a steel tube, which is topped with a pump. Turning a tap on in the house draws water from the storage tank, a float-operated switch in the tank goes on, the pump is set in motion, and water is pumped up to the storage tank. Figures suggest that the initial cost of the borehole can be recovered in under three years.

STORING RAINWATER
In rural districts, back in the 1930s and 40s, it was common practice to collect and save rainwater. The rainwater falls onto the roof, the water is collected and sent down a gully, the gully feeds the water over a filter bed, and water soaks down through the filtered bed into a tank where it is stored ready for use inside the house.

There is no denying that it is a big step to aim for a complete off-grid solution to water, but using stored rainwater for washing clothes, baths and flushing is a good midway option.

43

TOILET SYSTEMS

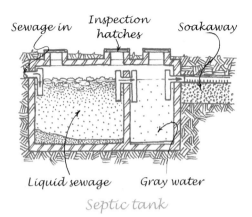

Sewage in   Inspection hatches   Soakaway

Liquid sewage   Gray water

Septic tank

When I was a kid in the 1950s and living in the country, going to 'the bog' or outhouse was a very primitive affair that involved walking outside to a little windowless building at the edge of the orchard, sitting over a great dark hole on a wooden shelf-like seat, spending time in the gloom while at the same time peering out through a line of little V-cuts that ran along the top of the door, doing the business, and then at the end of it all sieving a handful of dry earth or ash over my little excremental offerings. It was an ordinary everyday never-give-it-a-thought experience – no shiny white tiles or flushings of water, just lots of spiders, mice, strong smells, darkness and strange gurgling noises.

Every day or so when the bucket was full to overflowing, my grandfather would take it to the back end of the orchard, empty it into an open pit, and cover it with a few spadefuls of earth. When the pit was near full, he would fill it up with earth and then dig another pit. After about two years and 12 or so pits, he would go back to the first pit in line and dig out the contents – a beautiful brown crumbly mix, no smells or unpleasantness – and spread it over the vegetable garden. None of this seemed strange; it was just what we did.

Now, when we go to our beautiful shiny hygienic toilets, we do the business, and then flush it on its way with lots of very expensive top-quality drinking water. There is no denying that the development of the flush toilet is wonderful in the sense that we have done away with flies, rats, cholera and all the other problems associated with open sewers, but the issue that we now have is the vast amount of high-quality drinking water that we use for flushing. It is expensive and wasteful in terms of energy and resources. As to what happens to our excrement once we have flushed it on its way, much depends upon where you live. Not so long ago (in the 1990s) it was sent straight to the nearest convenient water source – a river or the sea – but now, more often than not, it is sent to a sewage works. If you live in the country, the chances are you have either a cesspit or a septic tank.

ECO OPTIONS

Using high-quality water to flush our excrement into the river or sea is a bad idea on many counts – it is a waste of quality water, it pollutes the river or sea, and the problem does not get any better. The sewage works is good, especially when it returns the end product to the land, but it is expensive in terms of transport and resources. There have been experiments with digesters – a process that turns sewage into gas – but really this only works on a grand town-size scale, or for say farmers who have mountains of manure to dispose of. The cesspit is no more than a pit that needs emptying, and it still needs the water flush. The septic tank solution is a good option, but only if you live in the country, and, while it still uses water, it could be gray water. Currently, one of the best options in the context of self-sufficiency is to use a 'Clivus' type system that turns the waste into compost – a reworking of my grandfather's idea of burying it for a couple of years and then putting it back onto the land.

CLIVUS

The Clivus (pronounced cleave-us) is a small-room-sized fiber and plastic box that is fitted with 3–4 pipes – two large pipes going up (one up to the toilet and the other up to the

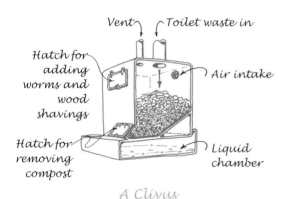

A Clivus

kitchen), sometimes another smaller pipe going to another sink, and a large pipe or flue going up and out at roof ridge level. You use the toilet or work in the kitchen and drop scraps of waste down a hatch. Either way, the waste falls into the enclosed container where it gradually breaks down. Very much like my grandfather's hole-in-the-ground scenario, after 1–2 years you can shovel out the most beautiful compost.

CLIVUS QUESTIONS

● **Does it smell?** No, for the most part. The downdraught from the toilet is so strong that the smells are drawn down from the kitchen and toilet and sent up the chimney.

● **Is it safe?** Yes. The process is such that the resultant compost is perfectly safe and user-friendly.

● **Does the compost look like excrement?** No. It breaks down to a fine crumbly golden brown tilth that good enough to spread on the garden.

● **Does the process need water?** No more than you would expect to get from urine and kitchen waste.

● **Can it be fitted anywhere?** No. Since there needs to be easy access to the emptying hatch, it needs to be fitted in a basement, or in a split-level house on a sloping site, or downstairs in a topsy-turvy two-story house where the toilet and the kitchen are upstairs.

● **Are there any problems?** Yes. Flies like the smell, but design modifications are sorting this by having extractor fans fitted to increase the updraught and discourage the flies.

● **What happens to the gray water from the house?** You can either put it down the public sewer, or, if space allows, drain it off into the garden. Another option is to send it to a reedbed.

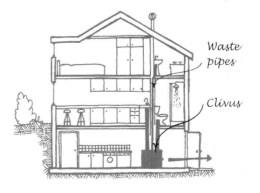

Clivus installation

SOLAR POWER

Solar power is term that we use to describe various methods of using the light and heat of the sun to provide us with energy. There are many ways that we can use to transform solar energy (see 'Solar options').

We all know about the heat from the sun; in very basic terms, it is hotter at noon than at sunrise or sunset, and it is warmer at the equator than at the poles. The amount of solar energy that reaches the earth – that reaches us – is determined by the angle at which sunlight strikes its target, and by heat absorption. If we wear black clothes we feel hot, and if we wear white clothes we feel cool. Black absorbs more heat – more heat from the wavelengths in the spectrum – than white. Some experts are of the opinion that dark matt green maximizes heat absorption better than matt black.

SOLAR OPTIONS
- Location – positioning our homes so that large windows face the sun at midday
- Passive storage masses – heat-absorbing walls of concrete and brick that absorb the sun's heat like a storage heater
- Shutters and curtains – these are used to hold in heat
- High-spec insulation in the walls, floors and ceilings
- Active storage mass solar panels – the sun heats a black painted surface and the resultant hot air is pumped around the building
- Active solar panels – fluid is heated and pumped into the heating system
- Photovoltaic cells – solar energy transfers into electricity
- Solar flues and chimneys – hot air creates cool air flow

PASSIVE DIRECT SOLAR GAIN

As regards using solar power to warm our homes, most of us are already doing it by passive direct solar gain. That is, we have shaped our houses so that they are in effect well-insulated boxes positioned so that there

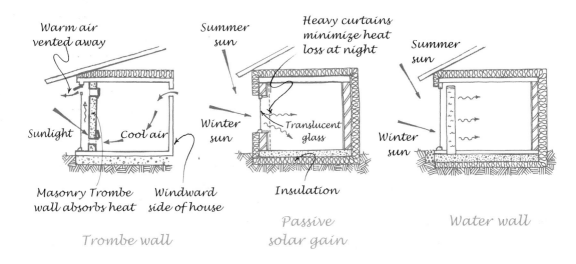

Warm air vented away

Sunlight   Cool air

Masonry Trombe wall absorbs heat   Windward side of house

Trombe wall

Summer sun

Winter sun

Heavy curtains minimize heat loss at night

Translucent glass

Insulation

Passive solar gain

Summer sun

Winter sun

Water wall

are large windows facing the sun at midday. The sunlight shines through the windows onto the various walls, floors and pieces of furniture within the house, where it is collected and stored as heat energy. If the night is cold, we draw the curtains or shutters to retain the heat that is gradually given off by the walls, floors and furniture – just like storage heaters – and if the night is hot we throw open the windows to let the heat out. If we take the direct gain idea one step further and have thicker walls and floors – with more masonry and concrete – more insulation within the floor, wall and ceiling cavities, larger areas of glass that are angled so that the rays of the sun will strike them at right angles, and use darker colors within the house, then the heat-storage capacity will be radically increased.

Whereas the passive direct gain approach is absolutely fine in a temperate climate where the days and nights are mild, it fails in a hot-cold climate (where the days are hot and the nights are cold) in that the rooms will be uninhabitable at midday.

PASSIVE INDIRECT SOLAR GAIN AND TROMBE WALLS

With indirect solar gain, a heat-absorbing wall of black-painted masonry or concrete, or a tank of fluid, is set between the window and the interior so that it receives the sunlight full on. With this system – known as a Trombe wall – the sun heats the wall, and vents in top and bottom of both the window and the wall are opened and closed to utilize the stored heat. The sun shines through the glazing and heats the masonry walls, with the effect that the space between the glazing and the wall becomes a thermal chimney, and then the vents that are set at floor and ceiling level in both the glass and the Trombe walls are managed so that the currents of hot air that rise by convection between the wall and the glass – in the thermal chimney – are directed either in or out of the building.

TROMBE CONTROLS

HOT DAY/HOT NIGHT In daytime the vents in the window are open and the vents in the wall closed. The air in the space rises by convection; hot air passes out of the two vents at the top of the window, drawing cool air into the bottom vent. The circulating air helps to cool the interior.

HOT DAY/VERY COLD NIGHT In daytime the top vent in the window and the bottom vent in the wall are open and the other vents closed. An additional window/vent towards the back of the interior is open. The air in the space rises by convection; hot air passes out of the top window vent and draws cool air through the interior space.

HOT DAY/COLD NIGHT In daytime the vents in the window are open and the vents in the wall closed. The air in the space rises by convection; hot air passes out of the two vents at the top of the window, drawing cool air into the bottom vent. At night the vents in the window are closed and the vents in the wall open. The hot air in the space between the wall and the window rises and passes into the interior.

COLD DAY/COLD NIGHT In daytime the vents in the window are closed and the vents in the wall open. The air in the space rises by convection; warm air passes in the vent at the top of the wall and heats the interior. At night all vents are closed.

SOLAR COLLECTORS

When we first became interested in self-sufficiency in the 1970s, people were experimenting with all manner of heat collectors – made from old central heating radiators, black plastic tube, rubber inner tubes, and so on. One system I remember was amazingly simple: water trickled down over a glass-covered galvanized steel roof into a trough and then on into a tank in the cellar, where it was used for hot water and space heating. This system did not look very pretty, and it was a huge free-standing structure almost as big as the side of a house, but for all that it was amazingly efficient, with cold water going in at the top end and too-hot-to-touch water coming out at the other. There was another system where black plastic pipe was wound round and round a massive cylindrical house-high storage tank. Just as before, cold water went in at one end and came out hot at the other. Commercial systems were coming onto the market at that time, but all in all they were bulky, very expensive, considered generally to be a bit alternative and hippy, and for the most part a bit hit and miss.

Now we have got to the happy state when commercially built solar collectors are not only available at a very reasonable cost – and the costs are coming down all the time – but moreover they are efficient, sophisticated and generally good, well-designed items. If you simply want to cut energy costs, and are looking for a tried and trusted method, solar collectors are a good option.

HOW DOES A SOLAR COLLECTOR WORK?

Although there are various compact and non-pump solar heating systems, most solar

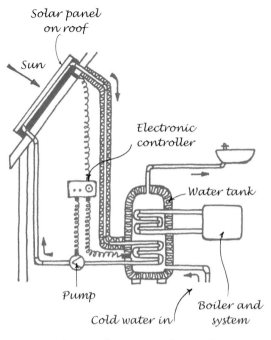

How solar panels work

collecting set-ups use one or more pumps to push the water/antifreeze/oil around the system. Solar heating systems are usually made up of five primary elements; a collector, a hot water storage cylinder, a batch of controls and sensors, one or more pumps, and pipework. This is how it works:

- The sun heats the heat absorber in the collector
- The heat from the absorber is transferred to the water/fluid
- At a fixed temperature the controls switch on and set the pump/s in action
- The hot water/fluid is pumped from the collector to a loop/heat exchanger in a storage tank
- The water from the storage tank is used either direct as hot water, or as space heating

TYPES OF COLLECTOR

There are three primary types of collector, all with slightly different qualities and characteristics. You need to research the options until you have a clear understanding as to which will best suit your needs.

FLAT COLLECTOR The collector is made up of a collection of radiator-like channels/tubes/pipes that sit on a thin heat-absorbing sheet within an insulated box and glazed box – very much like the old 1970s prototypes. The water/fluid in the tubes draws heat from the absorber.

WIDE-ANGLE CONCENTRATOR COLLECTOR

The collector is made up of a copper tube complete with fins that sit within a shaped collector housing – a bit like those old-fashioned bar electric fires. In many ways, this design is better than the flat collector in that the curved shape focuses the sun's rays on both the front and rear surfaces of the absorber plate.

EVACUATED TUBE COLLECTOR

The collector is made up of a series of transparent glass tubes that have contained within their body an inner and outer tube, a vacuum, a heat-absorbing surface, a mirrored heat-reflecting surface and a copper heating pipe. The sun's heat is absorbed by a coating on the inner glass surface; the heat passes to the tip of the heating pipe, and in turn is transferred to a copper manifold and then on to the storage tank.

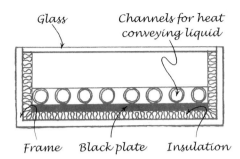

Glass · Channels for heat conveying liquid · Frame · Black plate · Insulation

Flat collector

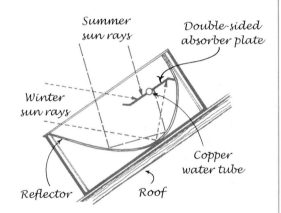

Summer sun rays · Double-sided absorber plate · Winter sun rays · Copper water tube · Reflector · Roof

Wide-angle concentrator collector

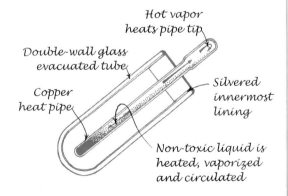

Hot vapor heats pipe tip · Double-wall glass evacuated tube · Copper heat pipe · Silvered innermost lining · Non-toxic liquid is heated, vaporized and circulated

Evacuated tube collector

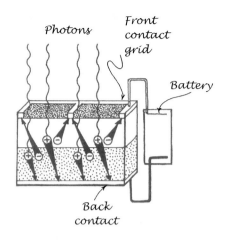

Photovoltaic cell

PHOTOVOLTAIC CELLS

In the late 1970s, photovoltaic cells (a spin-off from the space race and its associated industries) were developed as a means of directly converting light into electric power. No water – just sunlight directly into electricity. At that time, these cells were so incredibly expensive that really they were thought of as being beyond domestic use. Now of course we see photovoltaic panels everywhere – on top of telephone boxes, at the side of motorways when they are combined with little wind generators to provide small amounts of off-grid power for signs and the like, on the decks of yachts, and so on. Mass production has resulted in costs dropping rapidly. Whereas in the 70s costs were very high, they are now only a fraction of that price, and they are still falling.

HOW DO PHOTOVOLTAIC CELLS WORK?

A photovoltaic or PV system converts sunlight directly to DC electricity. A typical PV cell structure – made up in layers – has a back contact, two silicon layers, an anti-reflecting coating, a contact grid and an encapsulating surface. As to how it precisely works, it is enough to know that when the structure is bombarded by photons from the rays of the sun then a steady flow of electrons produces a minute amount of electricity in the form of a direct or DC current. Certainly, the amount that each cell produces is small, but if you connect a whole batch of cells in the form of a panel, and use an inverter to turn DC to AC, then you have a relatively simple and inexpensive energy source.

SOLAR COLLECTORS AND PHOTOVOLTAIC CELLS WORKING TOGETHER

Currently, there are many questions about PV systems. Will they come down in price? Are they reliable? How long will they last? Will they do the business in a cool climate? They are, however, starting to be used in a domestic context alongside the more tried and trusted solar collectors. All that happens, in effect, is that small photovoltaic panels are being used

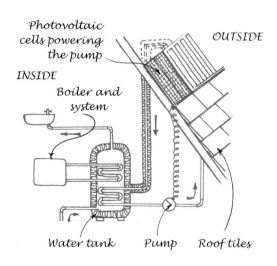

Pumped by the sun

to provide the electricity that sets the various pumps and controls in motion.

The system is not very sophisticated, and really in all fairness PV systems are capable of much more, but it is very neat. As production costs come down, so more products are coming onto the market.

PASSIVE INDIRECT SOLAR GAIN AND GREENHOUSE ROCK STORAGE

If you have a conservatory, you will know how quickly it heats up. They are great in late autumn and early spring – wonderful places to be. Even on a relatively cold autumn or spring day, they feel comfortable. Yet in late spring through to high summer and early autumn, when the sun is beating down, they are usually far too hot. For the most part, when the conservatory is too hot, say in early autumn and late spring, the house can be a bit chilly. At this point what most of us do, without really giving it much thought, is to open doors that link the house to the conservatory, so that the heat flows into the house.

You could take this basic approach one step further and build vents in the linking wall at floor and ceiling level – much like the Trombe wall – so as to create a chimney effect that draws hot air into the house. Alternatively, you could fit an extractor fan in the top vent and send hot air via ducting to some cold and distant part of the house. All you are doing, in essence, is taking the unwanted hot air and putting it in some place where it is wanted.

The 'greenhouse rock storage' option takes this way of thinking one step further, and in so doing very nicely tackles the house-too-cold-conservatory-too-hot scenario. The sun heats up the structure of the conservatory – the

walls and floors – so that the air becomes hot and starts to rise. At this point, an extractor fan switches on and draws the hot air out of the conservatory and on, via ducting, into an insulated space full of rocks, ideally an underground storage bin. Rock is usually the first choice, simply because it is low in cost and easily obtainable, but you could use any safe inert material that is slow to heat and equally slow to cool – such as crushed concrete, crushed glass, granulated metal. Once the heat from the air has been absorbed by the material in the storage bin, the air is circulated back to the conservatory as cool air. During the night, the hot air from the storage bin is sent by natural convection or by fan back to the house.

The rock storage approach is certainly more mechanical than the Trombe wall system, but it is a good option where there are problems associated with existing structures – meaning where you already have a greenhouse or conservatory that is too hot, and a house that is too cold. It cleverly tackles both problems.

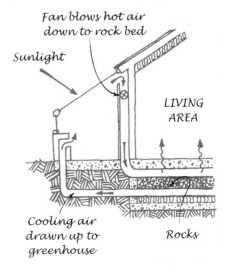

Fan blows hot air down to rock bed

Sunlight

LIVING AREA

Cooling air drawn up to greenhouse

Rocks

Rock storage

WIND POWER

It seemed obvious to us – there we were living in a place where the wind blew so fiercely that every tree and building seemed to be braced against it, so what was more natural than to build a wind generator? I wrote to a hippy commune, and was sent a batch of DIY designs telling me how I could make a generator using bits and pieces salvaged from old cars. When the big day came, the generator was mounted in place, the wires were connected, and the propeller began to spin. It looked so beautiful. To cut a long story short, we had power for a few glorious minutes and then the lights blew, the cables melted and the propeller cart wheeled off across the field. From then on, we settled for gas lamps and candles.

Now, of course, windmills are *de rigueur*. They are everywhere; rows of giant wind generators on hillsides and mountaintops, medium-sized machines outside petrol pumping stations, mini windmills alongside motorways. As one national newspaper headline so neatly put it, 'We are on the cusp of a Wind Revolution.'

The question now is not how to build one, or even where to get one, but rather what type of wind generator to install. In fact, there are so many options and so much jargon (e.g. utility tie-in, off-grid with battery bank, rooftop, traditional two-blade) that it can all be a bit baffling.

HOW DOES IT WORK?

We all know about traditional windmills. The wind blows, the sails go round, and the turning horizontal pivotal action is converted by means of gears, wheels and rods to vertical action to turn huge flat stones to grind corn, or to turn a pivot that turns a crank shaft that sets a water pump in motion. There are all sorts of complicated head and axis design options: canvas sails on a horizontal axis, cylindrical rotors on a vertical axis, vertical rotor blades on a vertical axis, blades like a child's windmill toy on a horizontal axis, and so on.

Old windmill

Most modern high-tech wind generators have anything from two to five blades or propellers of an aerofoil section on a horizontal axis – a bit like an old airplane. Most machines are mounted on a pole or tower, and designed in such a way that a mechanical or electronic governor comes into action and applies a brake when a top wind speed is reached. The wind blows over the blades causing them to rotate, the prop spins on a horizontal axis and turns a shaft, the shaft turns inside the generator, the generator converts the turning motion into electricity, and finally the resultant electricity is either stored in batteries and converted by means of an inverter to standard AC supply and/or fed into the national grid.

TECHNICAL QUESTIONS AND ANSWERS

- **How many blades?** Some machines have two, others three, and yet others five. There are so many designs that the only way forward is to read the data and choose what you consider to be a proven machine.

- **How much power can they produce?** There are many options: 0.6kW, 1kW, 1.5kW, 2.5kW, right on to 6kW and upwards. A 600W (0.6kW) machine will provide enough electricity to power lighting circuits in a three-bedroom house. A 2500W (2.5kW) will provide enough electricity to power lighting circuits and most appliances in a three-bedroom house.

- **How long will they last?** Much depends upon the machine, but one company claims that their machines have a product life of 20 years.

- **How high does the mast have to be?** Heights range from 18 ft for small machines right up to 50 ft.

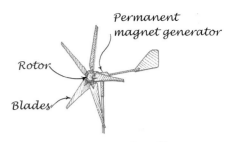

Permanent magnet generator

Rotor

Blades

Modern windmill

- **Can they be mounted on the roof?** Although the industry has been pushing for years for bigger hub diameters and taller and taller masts (based on the research that equates these figures with power output), there is now a push towards having lots of small machines rather than one big one. The idea is that every house could have one or more small rooftop machines – rather like most of us now have TV aerials or dishes. That said, there is an anti rooftop lobby which claims that vibration is always going to be an issue with rooftop designs.

- **Can I self-build?** The self-build option is a sensible idea if you are a good all-round mechanic with a fair knowledge of electricity.

- **What is best – grid-tie-in or stand-alone?** They are different; a small machine with batteries might be the only answer for an isolated rural location.

- **Are there any grants?** This depends on where you live. Good grants are available for Scotland, some parts of America, and Australia; poor grants for England and Wales. The situation is changing, however.

- **What are the planning issues?** If you live in a rural location and the mast is no more than 20 ft high, it should be fine, but always ask. You need planning permission if you live in a town.

53

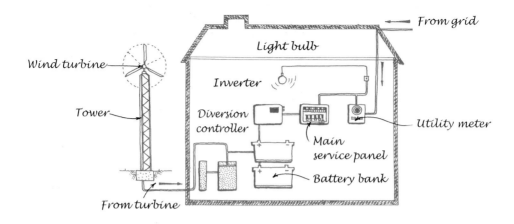

Wind electric system tied into the grid

OFF-GRID WIND POWER

An off-grid wind-powered system consists of a wind generator or turbine mounted on a tower/pole, a bank of deep cycle gel or carbon fiber type batteries, and an inverter. The wind blows over the blades causing them to rotate, the prop spins on a horizontal axis turning a shaft, the shaft turns inside the generator, and the generator converts the turning motion into electricity. The resultant electricity is stored in the batteries, and then passes through an inverter and on into the house as standard AC power. This is the perfect option – perhaps the only option – when you want to provide lighting for an isolated property. Another possible way forward is to leave out the batteries and use the electricity direct to power an immersion water-heating system, but of course this will not give you lighting.

GRID-TIE-IN WIND POWER

A grid-tie-in system consists of the wind generator mounted on a mast/pole, a utility tie-in inverter, a bank of batteries for a back-up system, a utility switch box, and a battery system switch box. The wind blows over the blades causing them to rotate, the prop spins

on a horizontal axis turning a shaft inside the generator, the generator converts the turning motion into electricity, and so on as with the off-grid system. The only difference this time – and the detail depends upon your choice of system – is that the electricity is either used first by the house with the excess being fed back into the grid, or it is fed straight into the grid, in which case the house takes power from the grid. Either way, you will have cut down on your total costs, used a renewable source rather than a fossil fuel, cut down on harmful emissions, and generally achieved a measure of independence.

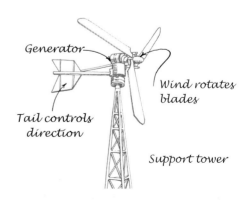

How it works

GLOSSARY OF WIND GENERATORS

AXIS The way that the turbine is mounted, and the plane on which it spins. Most modern wind generators spin on a horizontal axis; that is, they have a tail or vane at one end, a shaft and generator at the center, and rotating propeller hub at the other end. The tail swings round so that the propellers are facing into the wind.

BLADES The panels that are moved by the wind. Most small wind generators or turbines have three blades, but there are also two- and five-blade designs. Many experts agree that blade number and blade configuration are factors that relate to noise and vibration.

BRAKING The ability for the machine to cut down its speed, or close down its operation, if the wind reaches a dangerous speed. Older machines have some sort of mechanical centrifugal governor: weighted arms are thrown out by the spinning hub with the effect that brakes are applied. Most modern machines have some sort of furling brakes that are operated by the action of the propeller blades and/or the tail fin turning out of the wind.

JACOBS A company that has been around since the 1920s–30s producing one of the best wind generators. The overall structure of the windmill as illustrated on the left is referred to as a 'Jacobs' type.'

m/s Meters per second, referring to wind speed. 1 m/s = 2.24 mph (miles per hour). For example, if a machine is described as 'cutting in at 2.3 m/s,' it means it starts working at 2.3 times 2.24 mph, which is just over 5 mph.

PRAIRIE WINDMILLS Windmills designed to pump water, used by prairie farmers from the end of the nineteenth century to the 1950s. Some manufacturers produced little generators that could be fitted to the windmill pumps; they were just about powerful enough to light a few rooms. Many of the early wind generators – made in the 1960s–80s – were based on the multi-vane windmill designs.

ROOFTOP MACHINES In response to the need for renewable green energy for all, at least one company has produced a small rooftop-mounted generator that they claim is 'visually appealing, planning compliant, silent, and vibration free.' Some experts question the usefulness of rooftop machines, however; they are concerned about vibration, turbulence of the wind at roof level, and noise inside the building.

SAVONIOUS TURBINE A turbine that is made up of a series of curved, bow-like blades mounted on a vertical axis. The design is now thought to be inherently faulted.

TOWERS Most wind generators need to be mounted on top of a tower or mast – the bigger the machine, the higher, broader, stronger and generally more complex the design. The image of the turbine on top of the mast has become an iconic symbol that stands for clean energy. Generally, the higher the tower the faster and more constant the wind; low towers usually equate with low wind speed and high turbulence.

VIBRATION Shaking effect produced by wind generators. Vibration has always been a problem. Most wind generator failings are caused by a build-up of vibration associated with blade/propeller imbalance. One company that sells DIY machines stresses the importance of balancing the blades.

CAN YOU LIVE WITH A SMALL WIND TURBINE?

For many novices to self-sufficiency, the main goal is to live completely off-grid. For such people, the notion of having a small wind turbine is inspirational. Of course, there are people who live so far off the beaten track that they have no choice other than to produce their own electricity. If you are planning to go down the wind turbine road, you do have to face up to the fact that it can be a life-changing experience.

Basic three-blade turbine

OFF-GRID AND COMPLETELY INDEPENDENT

The scenario is that you live in an isolated farmhouse, the nearest power line is far away across the fields/hills/mountains/sea, it is windy for a good part of the year, you have plenty of wood for fuel, and you are aiming for total self-sufficiency. You are getting a small 1000W wind turbine.

- **Lights** No problem with the lighting if you fit a basic inverter so that you can run household power.
- **TV, radio and computer** You can run them via an inverter or call in a specialist to make adjustments so that they can be run on 12v. There might be some small amount of interference such as lights flickering and the radio squeaking.

- **Electric kettle** You cannot run a kettle; it needs too much power.
- **Heating and hot water** Yes and no. You could dedicate the wind turbine to heating the water via a low-voltage immersion heater, but then you would not have the lighting. You could store the power in a bank of batteries and use a water immersion heater to take care of the 'overflow' power meaning the excess 'dump load.'
- **Electric cooker** You cannot run an electric cooker; it would need far too much power.
- **Washing machine and dishwasher** Yes, if you have a good-sized bank of batteries and are prepared to run the machines at carefully chosen times on economy settings. In many ways, however, it would be much easier to do without the washing machine and the dishwasher and go for a woodburning stove and a washtub. We have done it!
- **DIY power tools** Much depends upon the tools. Why not use hand tools?

GRID TIE-IN AND COMMITTED TO SAVING ENERGY

The scenario in this case is that you live in a large house in the suburbs with a large yard, and you are aiming for near self-sufficiency without actually cutting yourself off from the main electricity grid. You have lots of time, but limited funds, and you want to cut your use of the grid back to the bone. You have planning permission for a small 1000W wind turbine. Apart from the turbine, you want to live much as the first inhabitants of the house did back in the 1920s.

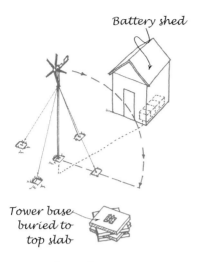

Wind turbine set-up

- **Lights** No problem with the lighting. A small wind turbine should give you more than enough. On a windy winter's evening, you could cut electricity costs to nil simply by turning out the lights and sitting around a multifuel stove. It is good fun – warm toes, romantic firelight, making toast … just like the first inhabitants of the house.
- **TV and radio** You can run these appliances just as you always did. The only thing is that, if you have been happily working out in the vegetable garden, sorting out the chickens, handwashing clothes and so on, the likelihood is that you won't want to watch the box.
- **Computer** You can run it just as before.
- **Electric kettle** You can run an electric kettle just as before, as long as you are aware that most of the power will come from the grid. Why not put a kettle on the stove instead?
- **Heating and hot water** You can use an electric immersion heater just as before, but a good part of the power will come from the grid. A multi-fuel stove with the back boiler is a better option. Another way

forward is to think hard about how you use hot water and then act accordingly. For example, with a bath don't heat the water to near boiling point and then add cold to cool it down to a safe temperature. It is better to settle for just heating it to a safe washing temperature.

- **Electric cooker** You can run a cooker, but just be aware that it will use a lot of power. My Welsh granny used to cut costs by eating lots of cold or easy-cook meals. It sounds miserable, but far from it – fresh bread, local cheese, lightly steamed fresh vegetables, salads, fresh fish that was cooked in a minute … I really don't think we suffered!
- **Washing machine and dishwasher** You can run a washing machine and a dishwasher just as before, but why bother? The dishwasher is a waste of space, and you could go for a woodburning stove and a washtub. You will have to spend time sloshing about with tubs, buckets, a small mangle, a washing line and lots of washing airing around the fire, but it is not such hard work. Think about it. Do you really have to wash your shirt after one wearing? Do you get so sweaty that every stitch needs to be washed every day? You could cut back costs/energy and save on all sorts of pollution problems – too much soap, too much laundry detergent – simply by doing a small handwash.
- **DIY power tools** Traditionally, most people made just about everything with a kit of carefully chosen hand tools. They were not bothered by electric cables, potentially dangerous fine dust, or the high-pitched sound of a power tool. So, try to cut out power tools altogether, and use good-quality hand tools instead.

GEOTHERMAL HEATING

In temperate climates, while outdoor temperatures go up and down with the seasons, temperatures below ground remain more or less constant. For example, in Scotland, in mines and holes 330 ft deep the temperature of underground water stays constant at about 54°F. At 660 ft, the temperature goes up to 59°F. The deeper you go, the hotter it gets.

A geothermal system, in the form of a flexible plastic pipe loop filled with a water/antifreeze mix, with the whole loop running underground, takes advantage of these constant underground temperatures by absorbing the stored heat and carrying it into the house. In winter, the system within the house extracts and compresses the absorbed heat from the water/antifreeze loop and distributes it throughout the building as heat. In summer, the same system reverses the procedure to cool the building.

Buried below the surface

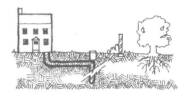

*Drilling down
on a difficult site*

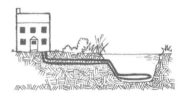

Using a large pond/lake

SYSTEM OPTIONS

- **Horizontal loop** This is a good choice if you have plenty of land. A 330–660 ft length of pipe – the loop – is buried in trenches or a block excavation at a depth of about 5–6 ft. This is a low-cost option if you are starting with a clear site, and/or if you are able to do some part of the trench/ground work yourself.
- **Vertical loop** This is a good choice if you only have a small plot of land. The pipe loop is set in a small-diameter borehole at a depth of 130–500 ft. Although this is a good option in terms of space, boreholes are undoubtedly more expensive than trenches.

- **Water loop** This is the perfect choice if you have a good-sized pond or lake. The loop is simply set in place under the water and the task is complete.

WHAT YOU NEED

The total system is made up of the heat pump, the heat-exchange medium – the liquid in the enclosed pipe loop in the trench borehole or under the water – the pipe or ductwork within the house, and of course electricity for the various pumps. The main question that you need to ask yourself if you are thinking about installing such a system is where to put the loop. No problem if you have a large lake in your backyard, but for most people the choice is between a borehole or a pattern of trenches. On the face of it a pattern of trenches or a swimming-pool-sized excavation might seem to be an easier option than a borehole, but much depends upon the nature of the subsoil. If the soil is very loose and sandy, or made up of bedrock or waterlogged clay, then it might well mean that you have no choice other than to go for the borehole. Then again, as excavating technology – trenches and pools – is relatively commonplace and widely understood, the likelihood is that it could be achieved using low-cost local labor.

USING THE HEAT

Once the heat has been extracted from the loop, it can be used in the form of hot air or hot water. For the most part, hot water is

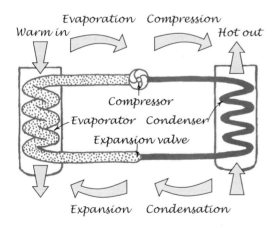

Geothermal air conditioner and boiler

more useful in that it can be used both for central heating and for the domestic hot water supply – for washing, baths and so on. With an average system, the warm water from the loop is passed through a heat pump where the temperature is raised to about 120°F.

This water is stored in a large, well-insulated storage tank. For space heating, the water from the storage tank is simply used to heat radiators. For a domestic hot water supply, the water from the storage tank is pumped through indirect coils in a hot water cylinder and used for washing and bathing.

HOW EFFICIENT?

The best way of working out the efficiency is to compare geothermal with fossil fuels. Geothermal heat pumps are rated according to their COP (coefficient of performance), that is according to how much energy is used set against how much energy is produced. Geothermal systems are very efficient in that for every unit of energy used to power the system about four units are supplied as heat. To put it another way, while coal is about 70–90 percent efficient, a heat pump is 400 percent efficient.

BOREHOLES

Boreholes are interesting in that they can be used to supply both water and heat. If you are going to the trouble and expense of having a hole sunk for a geothermal system, then you could also use it to supply you with water. You would finish up with a borehole with two pipework systems in place – a sealed loop for the thermal heating, and a pipe with a submersible pump for the water. Keep this in mind when you are sinking the borehole.

MAINTENANCE

Geothermal systems are relatively easy to maintain. The pump and compressor will need regular check-ups but, apart from that, the sealed loop in the borehole or trench and the ducting or pipework are more or less maintenance-free.

GEOTHERMAL QUESTIONS AND ANSWERS

- **What about costs?** Much depends upon your particular situation. If the soil conditions are right, and if you can use local labor to excavate a swimming-pool-sized hole or cut deep trenches, then you will be able to cut costs to the minimum.
- **What does a borehole involve?** We had a borehole dug for a first home, it was noisy and expensive, and it made a terrible mess in the garden. That said, costs have come down and the equipment is more efficient.
- **Are there many borehole drillers out there?** No problem in places like the USA and Australia, but not so easy in the UK. I have spent two days on the internet and phoning around and I still have not found one!
- **Is geothermal heating an easy option?** This depends on where you live. It is fine if you live in the USA, Canada, Australia or Japan, or just about any place where there is a shortage of fossil fuels and/or a long tradition of sinking boreholes, but not at all easy in the UK and some parts of Europe. The good news is that many governments are positively advocating the geothermal option in the form of grants and tax cuts.
- **Is it a cutting-edge technology?** No, geothermal technology has been around and understood at least since the 1930s–40s. The problem has always been with the sealed loop – a difficult and expensive option when the pipes were made of steel or copper. The good news is that the introduction of low-cost plastics has brought the cost down.

WATER POWER

MODERN HYDRO TURBINES

Small-scale water power is a big issue, with 'eco' groups, governments and manufacturers all beginning at long last to wake up to the needs and possibilities. One such manufacturer, Pico, based in Nepal, has come up with a revolutionary hydro turbine that is low in cost, easy to maintain, and above all compact and mobile. It looks more like a large egg whisk than the usual sort of Mississippi steamboat paddle-wheel shape that most of us have in our mind's eye. In effect, it is just a shaft with propellers on one end and an alternator on the other, with the whole thing mounted within a frame.

The turbine is carried to the nearest stream, arranged so that the propeller end is suspended under a torrent of fast-flowing water, and then wired up to run a few lights. Although this turbine is designed for third-world communities – in situations where there is plenty of water, a need for a small amount of electricity, and very little engineering know-how – it is plain to see that it would fit very nicely into a self-sufficiency scenario. Technically, given a 5 ft head of water, and a water flow of about 19 gallons per minute, this machine is able to produce a capacity of 500W. Put another way, if you set it up in a fast-flowing stream and build a containment, sluice or slide to direct the water so that it can gush straight down onto the propellers, it will provide a small but reliable supply of electricity – perfect for your lighting, radio, TV and computers and any other low-powered equipment.

SELF-SUFFICIENCY AND WATER POWER

If you do have access to water, and you have a legal right to use it, and you have checked out the depth, the possible drop height, the rate of flow, and whether or not it flows in summer months, you then have to consider just what it is that you want to power. The likelihood is that, unless you want to grind corn or power some part of a mechanical process such as pumping water, hammering iron, running a loom or mixing clay, you will want to run a water-wheel to produce electricity. As for how much electricity you can produce, this obviously relates to your particular set-up. If you have actually moved into an old watermill, one with a huge undershot wheel, then you can simply mount a turbine on the shaft and produce as much power as you need.

For most people, however, we are back to the small stream set-up. With this scenario, you are faced with much the same factors as with a wind turbine; there is a higher power potential in the winter than in the summer. This is fine, because the likelihood is that you will need more power in the winter but, much as with the wind turbine, you are also faced with having periods when there is so much power potential that there is a need to store it in a bank of batteries.

The best advice is to look at your situation on the ground, consider your needs and then sort out a water turbine to suit. Our findings confirm that in this instance a small turbine in a large stream is an easier option than a large turbine in a river. The reason for this is that a large river scenario is difficult in that not only are you more likely to have problems getting water rights, but the engineering problems involved in building a suitable set-up can be truly monumental.

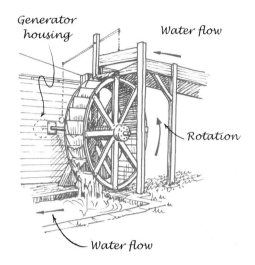

Traditional waterwheel attached to an electrical generator

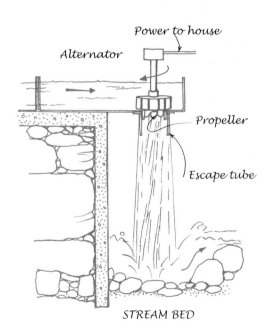

Pico-hydro turbine - a modern low-cost, easy-to-maintain water turbine

RECYCLING

Not so long ago our everyday household rubbish – a few brown paper bags and newspapers, a breakfast cereal carton or two, a couple of empty tin cans, and a small amount of kitchen waste – was either composted or recycled. On the down side, however, almost every household had a coal fire and a kitchen boiler belching out black smoke, every factory had coal or coke boilers, and all our raw sewage was dumped in the nearest lake or ocean.

Now there is a clean-up operation, solid fuel fires are controlled, sewage plants are more sophisticated, and seas and rivers are being cleaned. In many ways, there have been massive improvements. On the down side, most households now throw out about five times more waste than they did back in the 1950s.

The average waste bin contains: paper 30 percent, wood 13 percent, garden waste 10 percent, metals 9 percent, miscellaneous non-organics 8 percent, textiles 8 percent, plastics 7 percent, miscellaneous organics 6 percent, food waste 6 percent, and glass 3 percent. Of all this waste, we recycle: newspaper 80 percent, cardboard 70 percent, steel cans 60 percent, garden waste 56 percent, aluminum cans 43 percent, car tires 35 percent, magazines 33 percent, plastic water bottles 31 percent, plastic soft drink bottles 25 percent, and glass containers 22 percent.

If you look at the other side of the recycling figures, however, you will see that the fact is that a growing amount is going into landfill. For example, if the figures say that 35 percent of car tires are recycled, then it follows that a massive 65 percent of tires are piling up or being put into landfill sites.

WHAT WE CAN AND CANNOT DO AS INDIVIDUALS

- We cannot go back to the good old days of the 1950s when we did not have plastic packaging, but we can try to reduce the amount of packaging as far as possible either by growing our own food or by buying local produce.
- The best approach to managing waste is to avoid creating it in the first place, so we can each try to reduce the amount we produce by reusing both containers and products.

Most waste could be recycled

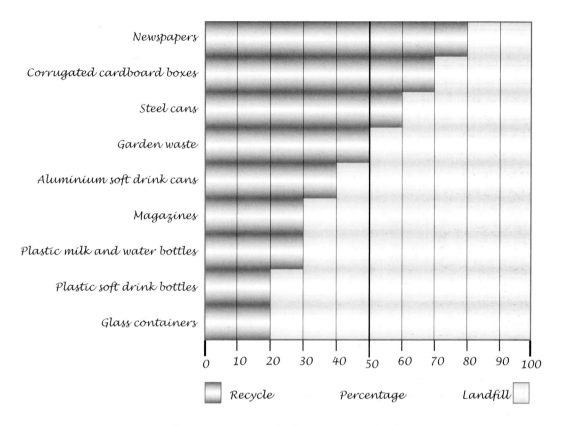

What is currently being recycled

- Recycling saves energy. For example, recycling aluminium drinks cans saves 95 percent of the energy that goes to make the cans from a virgin source – less transport, fewer raw materials and less pollution. If we give the canned drinks a complete miss, however, we could save energy on all fronts.
- Plastic carrier bags are a menace, and we don't need them. We could use our own bags, and we could lobby and worry for the reduction of plastic bags, or at the least for the introduction of biodegradable options.

- In the context of self-sufficiency and recycling, our mantra must be 'reduce and reuse, reduce and reuse.' Growing our own food not only cuts back on packaging, but more than anything it cuts back on the energy involved in transport. For example, if we grow our own food and consequently eat food that is in season, it follows that we will be eating local products rather than food that was grown on the other side of the world. This all adds up to less global transport and less packaging.

INSULATION

Heating and cooling our home accounts for about 75 percent of our energy costs. In an older-style, totally uninsulated house, the heat leaks out almost as fast as it is produced through gaps and cracks in the structure, and through materials that conduct heat. Figures suggest that 25 percent of the heat goes straight up and out through the ceiling and roof, 35 percent through the walls, and 15 percent through the floor.

R-VALUES

The more efficient your insulation is, the slower the heat loss will be. The quality of insulation – meaning its thermal resistance – is measured or rated in terms of 'R.' The better the insulation, the higher the R-value. Our aim should be to push the R-value as high as it will sensibly go. Put another way, a high R-value = low heat loss = low energy consumption = minimum energy costs = saving on global energy resources = all good.

INSULATION AND TRADITIONAL HOUSES

Traditional structures have always been relatively well insulated – caves, thatched cottages, thick stone walls built with a lot of rubble and earth infill, hugely thick cob walls made of mud and straw, timber-framed buildings with the cavities stuffed full of straw – they are all comfortable in terms of heat loss. For thousands of years, people used local materials to build their own houses. Our ancestors would not have known anything about design and R-values; they just knew that walls had to be low and thick in order to take the weight of the roof. For example, a

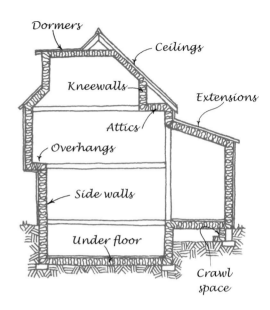

Insulation in a house

traditional Scottish croft house has walls about 2–4 ft thick and a thatched roof that more or less touches the ground. Such houses would not have been very hygienic – with the livestock living in the adjoining rooms, and the low rooms full of peat smoke – but they were warm in winter and cool in summer.

The Maori people in early nineteenth-century New Zealand were also very comfortable, living in beautifully crafted buildings made of carved wood, mud, stone and thatch; they were warm and dry in winter, cool and dry in summer, they could not hear the rain, and the whole building was sustainable in terms of construction materials. Our ancestors learned by trial and error over countless generations that, if they built their homes in a certain way, this desired effect would be achieved.

INSULATION AND MODERN HOUSES

The Industrial Revolution put paid to thousands of years of intuitive building when it became possible to build houses with thin walls and roofs – houses that were built by factories rather than families. It could not have been much fun for my grandfather – a coal miner – who lived in a small terraced house in Wales. It was a little 'modern' house with paper-thin walls, and a thin Welsh-slate roof. The coal heap was literally as big as the back garden, and yet it was always cold and draughty. Much the same could be said for the New Zealand Maoris in the 1890s when they lived in their new 'improved' corrugated-iron kit huts. So it was from the middle of the nineteenth century right through to about the 1930s – houses were built with little or no regard to traditional methods.

Around the end of the second world war, however, designers gradually came to realize that insulation techniques could be used to make homes more comfortable. At first, they used straw, minced-up newspapers, rags … just about any low-cost byproduct. Insulation as we now recognize it was first introduced around the 1940s. At that time, it was tentatively suggested that the ideal was to insulate with such materials as plasterboard, fiberboard on battens, glass silk – probably glass fiber – cork board, slag wool and aluminum foil. Yet, builders being builders, and profit margins being all-important, nothing much happened on the insulation front until the 1970s, when the oil crisis forced energy prices to climb. Interestingly, while we now all know about the importance of insulation, most professional and self-builders are still opting for the minimum thickness, rather than trying to achieve maximum insulation.

TYPES OF INSULATION

The problem with insulation products is that, while, for example, foil-foam-foil is a top-notch product, it is not natural or particularly 'green.' Sheep's wool is a natural product, but it is expensive and needs to be used in thick layers, which affects transport costs, and so on. The difficulty is that you cannot simply compare one product with another: you must look at the whole gamut, and then choose what you consider is the best middle way.

FOIL-BUBBLE-FOIL (FBF) Made from a sandwich of aluminum foil–polyethylene sheet–polyethylene bubble–polyethylene sheet–aluminum foil. At $3/16$ in. thick, it has an R-value of R-14.1.

FOIL-FIBERGLASS-VINYL (FFGV OR FFGV) Made from a sandwich of aluminum foil–polyethylene sheet–fiberglass–polyethylene–reinforced vinyl. At $1/4$ in. thick, it has an R-value of R-10.3.

FOIL-FOAM-FOIL (FFMF OR FFMF) Made from a sandwich of aluminum foil–polyethylene sheet–close-cell poly–fiberglass– poly-ethylene–aluminum foil. At $1/4$ in. thick, it has an R-value of R-14.5.

WOOL Natural sheep's fleece bonded with natural rubber. It has an R-value that goes up according to its thickness; for example, at just over 6 in thick it has an R-value of R-23. A completely natural, sustainable product – reusable, biodegrad-able, no toxic gases, not irritating to the skin, good for new build or remedial DIY, and it comes in thicknesses right up to 1 ft.

INSULATING A TRADITIONAL COLD ROOF

The accepted norm with good-quality housing built between the late 1950s and say the late 1980s was to place fiberglass insulation into the cavity as the building work progressed, and then when the building was finished to lay fiberglass insulation in the attic space. This was a great improvement on what went before, and it looks good on the drawing board – very trim the way the insulation sits there as a nice neat layer and tucks down between the ceiling joists and rafters – but in reality its installation is a bit of a problem. It is a problem on two counts: the wad of insulation between the rafters restricts the air flow, and it is tricky to install.

The task of actually putting the fiberglass into the sharp angle between the rafters and the ceiling joists is both difficult and unpleasant. If you look in attics of this period, you will see that not only is the sharp angle between rafters and joists usually uncovered but, worse, the insulation in the cavity has settled and slumped. All this adds up to a poorly insulated area that usually shows itself in the rooms as a damp or stained patch of condensation on the junction between walls and ceiling. This can be a big problem with bathrooms and kitchens that are situated on the cold side of the house.

INSULATING A HIGH-QUALITY DORMER ROOF

The good thing about roof insulation is that it can, to a great extent, be installed as a remedial DIY task at a post-build stage. With the scenario as illustrated, the area below the roof is intended to be a high-quality living space – kitchen, bathroom or bedroom, for

example. From inside to out, the order of materials is: plasterboard, insulation between counter-battens, a vapor barrier, insulation between rafters, a breather membrane, battens to hold the membrane in place, counter-battens for the slates, and finally the slates.

If you are living in a traditional cold-roofed house, you can greatly improve the insulation by stuffing the space between the rafters with insulation, and then following it up with a vapor barrier, counter-battens, more insulation, and finally the plasterboard or tongue-and-groove wood. The vapor barrier on the underside of the rafters prevents the moisture produced by all the normal household activities – breathing, cooking and cleaning – from passing through the ceiling.

If you are short of cash, and/or you want to complete the task in several stages, start with the insulation between the rafters and then follow on with the vapor barrier and the plasterboard. At a later stage, when money and time permit, you can add the counter-battens and the additional insulation and surface.

INSULATING A DORMER WARM ROOF

While the main object of the exercise is to achieve a warm, well-insulated roof space, the space is also good for occasional use. Its insulating characteristics are far better than the old cold roof – good if you are worried about water pipes and frost – but not so good that you can use it for a kitchen or a bathroom. If you look at the drawing, you will see that from inside to out the order of materials is: plasterboard, insulation stuffed into the space between the rafters, breathable membrane or roof felt, battens and tiles. This arrangement is good in that the contained insulation allows the roof space to be used, but not as good as it could be. The insulation is still minimal and there is no internal vapor barrier.

If there are suitable stairs, and adequate lighting and ventilation, this arrangement would be fine for use as an office or an occasional guest bedroom.

INSULATING A TIMBER-FRAME WITH BRICK SKIN

Timber-frame is a building technique that has its roots in structures like medieval cruck-framed houses, wattle-and-daub huts, and frame-and-weatherboard fisherman's cottages. At its most basic, it is a wooden frame with the spaces infilled with wattle and daub. In the eighteenth century, when brick houses were considered to be high-status, a technique developed that involved covering existing timber-frame houses with a brick skin. While such buildings were undoubtedly warmer and drier, the technique came to be associated with concealment.

In the UK at least, timber-frame has always been less popular than bricks and mortar. This being so, while timber-frame was always considered to be fine in, say, the USA – they are happy with the wood being on show – when it comes to the UK and Australia, it is generally concealed.

If you look at the drawing on the right, you will see that, not only does modern timber-frame construction offer a good opportunity to insulate at the initial building stage, but the structure is also such that it can be easily upgraded at a later stage, simply by adding another layer of insulation to the inner skin.

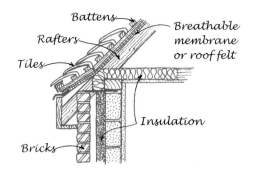

Traditional roof insulation

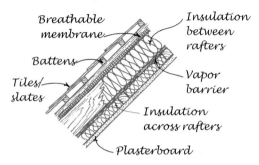

Roof insulation

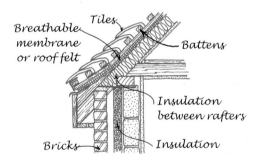

Dormer roof insulation

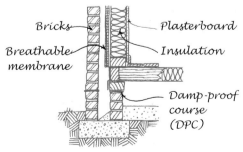

Walls: timber and brick

67

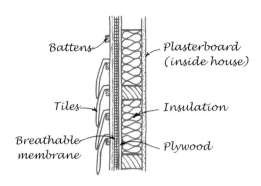

Wall: timber/hung tiles

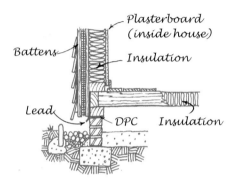

Wall: timber/weatherboard

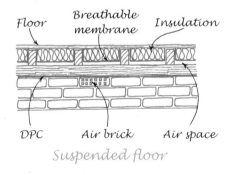

Suspended floor

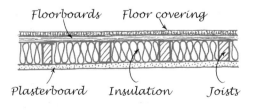

Intermediate floor

INSULATING A TIMBER-FRAME WITH HUNG TILES

Traditionally timber-frame with hung tiles was considered to be the quality option for the upper story or first floor of a brick or stone building. Before the 1950s, the tiles were either pegged-and-hung – meaning two wooden pegs were pushed through holes at the top of the tile so that the tile could be literally hung on the batten – or the timber-frame was clad in tongue and groove before being battened and tiled. If there was a waterproof membrane (usually not), it was either tarred paper or felt.

The order of materials from inside to out is: plasterboard, the timber-frame stuffed full of insulation, shuttering plywood spiked to the outside of the frame, a breathable membrane, vertical battens, horizontal counter-battens, and the tiles. If the tiles are in good condition, you can work from the inside and remove the plasterboard, and then stuff the cavity with insulation and refit new plasterboard, or you can leave the plasterboard in place, counter-batten on top of the plasterboard and fit one of the high-tech foil-bubble-foil insulation options. If the tiles are in poor condition, you could tackle the problem from the outside: remove the tiles and battens, stuff the cavity full of insulation, fit a breather membrane, fit vertical battens, and then counter-batten and fit the tiles.

INSULATING A TIMBER-FRAME WITH WEATHERBOARD

If you live in a single-story, pre-1950s, timber-frame cabin, the chances are that it has a very basic frame structure. The order of materials from inside to out is: plasterboard or plaster and lathe, the timber-frame stuffed with insulation, plywood or battens, tarred paper or felt, vertical battens, and the clapboard, tongue and groove or weatherboarding. Structures of this character are pretty rudimentary, but the advantage is that they are a joy to work on.

When we came to sort out our little cottage, which was just like this, we found three slightly different scenarios: a part built in about the 1920s that was insulated with wood shavings with tongue-and-groove boards on both sides; a part with insulation made from a sort of mattress of shredded paper (it was a commercial product); and a part that was totally uninsulated apart from having a product called 'Beaver Board' pinned on the inside. In the end we removed the clapboarding, stuffed the cavity with sheep's-wool insulation, fitted a breather membrane, and refitted new clapboarding – all from the outside; then we went inside, counter-battened over the plasterboard and fitted high-tech, foil-bubble-foil insulation topped off with tongue-and-groove boarding. We also removed and replaced the floorboards, and fitted insulation between the joists.

INSULATING A TIMBER-FRAME WITH SUSPENDED WOODEN FLOOR

If you live in a pre-1950s house, the probability is that it has a suspended wooden floor, meaning boards spiked to joists. Such floors are usually uninsulated. If you lift a floorboard you will see either joists sitting on low walls, or joists that are hung off a wall plate. There are two relatively low-cost options: you can lay a plywood-foam-plywood flooring directly on the old, and top it off with a floor covering of your choice; or you can carefully lift the original floorboards, staple a heavyweight plastic plant netting across the joists, lay your chosen insulation between the joists – so that it is

supported in a sort of netting hammock – and then replace the boards. There are many clips, and foam-and-clip systems, designed to sort out this problem, but they are very expensive and rely on the joists being placed at very specific centers – not a good option when you are dealing with older buildings.

INSULATING AN INTERMEDIATE FLOOR

An intermediate timber floor is that one that you walk on if you go upstairs, usually the floor of a bedroom or attic. In most domestic buildings, a traditionally built intermediate floor is just like a suspended ground floor – the joists are set at centers and hung from wall plates. The only difference, of course, is that the plasterboard or tongue and groove on the underside of such a floor is seen as a ceiling. The two primary reasons for insulating an intermediate floor are to reduce sound and to stop heat from traveling upwards.

Say, for example, that you want to insulate an intermediate floor, because you would like to use the attic for storing apples. It needs to be cold but frost-proof. Much depends on the condition of the existing floor and ceiling. If the floor is sound, then simply lift the boards, set the insulation in place, and refit the boards. Much the same applies with the ceiling: remove the plaster, stuff the cavity full of insulation and fit a new ceiling. The only difference with the ceiling is that the insulation has a tendency to fall down under its own weight, so you need to hold it in place while you are fitting the plaster. When we repaired our ceiling, we considered the various options, but in the end we settled for using slabs of sheep's wool. We found that if we cut it slightly oversize it was firm enough to wedge between the joists and stay put for the time it took to fit the plasterboard.

THE ORGANIC
FOOD GARDEN

ORGANIC GARDENING

Organic garden design

Pesticides and other chemicals are present in the food we buy but it is not known how toxic these are in the long term. Many farmers believe that the only sensible way forward is to say no to chemicals, pesticides and artificial fertilizers, and yes to good-quality plant and animal manures, and to respecting nature. Perhaps the days of farmers indiscriminately spraying pesticides and herbicides, battery egg producers feeding their hens a mix of recycled chicken manure and antibiotics and cattle farmers feeding recycled abattoir waste to their stock, are numbered.

The good news is that most people appreciate that organic farming can be equated with tastier food, healthier eating, more exercise, improved nutrition, healthier environment, better soil conditions and better wildlife. One country after another is declaring that the future will be organic.

PLANNING AN ORGANIC SELF-SUFFICIENT GARDEN

Try likening your garden, allotment or smallholding to Robinson Crusoe's desert island. Your garden is your island, you have gathered and salvaged all your tools and essential supplies, you have built the infrastructure – all the beds and raised borders, a few chicken runs, a compost heap and a polytunnel – you have an endless source of animal manure, you have an inexhaustible coming in and no waste going out. There will be difficulties, and there may be times when you have to compromise using plastic sheeting, wire fencing, fleece or plastic netting, for example, but the trick is just to do your best with what is available.

HINTS AND TIPS

RECYCLE Use, reuse and recycle everything from weeds, rainwater and newspapers to ashes and food waste.

ANIMAL MANURE This is the best fertilizer you can get. Keeping chickens is good, as you get manure, eggs and meat.

PLASTICS Plastics, in the form of sheeting for polytunnels and cloches, netting, plastic plant pots and so on, are in many ways a necessary evil. The problem is that they are made from petrochemicals. They are not poisoning our back yards, but in the grand scheme of things they are poisoning our environment. There are natural alternatives – glass, cotton netting, clay pots – but they use energy. You will just have to think it through, weigh the costs and act accordingly.

WEEDS Weeds are good! You can feed them to the chickens, turn them into compost, eat some of them, turn them into mulch, use them to attract wildlife; there are lots of possibilities.

COMPOST Turn your waste into compost and feed it into the soil.

ENVIRONMENTAL IMPLICATIONS Think carefully about 'hard landscaping,' such as fences, gates and paths. Stone, greenwood and earth are better than wire, plastic and concrete. For example, a path made from chipped and shredded wood not only looks good, and encourages all manner of insects and birds, but once it has broken down it can be fed back into the garden. A wire fence around the chicken pen will eventually break down and rot back into the soil, unlike a plastic fence.

BONFIRES I love a good bonfire, but considering the smoke and the gases it is really better to try to turn it all into compost. Make heaps of twigs and branches – they will rot down and encourage insects, snakes, mice and birds.

WATER Collect rainwater in butts and ponds. Gray water – from baths and showers – can be used to water the garden, as long as it is not overloaded with soap and other cleaning agents.

ANIMAL PESTS If you have lots of rabbits, mice, squirrels or similar in your garden, and you think of them as pests, get a cat or dog. Perhaps you could eat the rabbits.

NATURAL PEST CONTROL Once you have established a clean environment, with no chemicals, the organic 'wheel' will begin to turn. Big insects will feed on small insects, birds will eat insects and snails, chickens will eat just about everything, toads, frogs and newts will do their thing, and so on. Pests – the creatures that you perceive as being pests – will simply be eaten up and become part of the great scheme of things.

LOCAL SUPPLIES If you do have to bring in supplies, at least do your best to get them locally. Encourage local crafts, rather than buying in products that have traveled halfway around the world.

COMMON WILD FOOD

Warning Always choose wild food with care. If you do not recognize a nut, fungus or other possible food, ask the advice of a local expert, or steer clear of it.

Agaricus arvesis **(Horse Mushroom)** Grows in meadows and pastures. Often as big as an outstretched hand, with a large cap and pinky-gray to brown gills. Mature examples flatten out with the gills turning a dark meaty brown. One good-sized mushroom will fill a large frying pan to overflowing – easily enough for two greedy people. Very tasty fried whole and layered with a crisp-fried omelette and thin brown bread to make a sandwich – bread-omelette-mushroom-omelette-bread – all washed down with cider!

Armoracia rusticana **(Horseradish)** Grows like a weed in many parts of the world and is common on rough, neglected ground. Easily recognized by its long, crinkle-edged leaves and long, parsnip-like roots. Horseradish sauce is perfect with roast beef or barbecued fresh tuna steaks, or spread thinly over cheddar cheese and made into a sandwich.

Cantharellus cibarius **(Chanterelle)** A yolk-yellow, funnel-shaped fungus with fan-vaulting gills on the underside, a frilly edge to the cap, and no ring around the base, commonly found in woodlands from later summer through to autumn. Much favored in some parts of France and the UK, where it is eaten with scrambled egg and toast.

Castanea sativa **(Sweet Chestnut; Spanish or Prickly Nut)** The nuts, which are produced by a good-sized tree, are enclosed in a very prickly shell and cannot be mistaken. They can be gathered in autumn. I like them raw, but Gill prefers them boiled and mashed with a knob of butter and greens on the side, or roasted on an open fire and dunked in olive oil. If you come across a bumper crop, they can be pickled.

Conopodium majus **(Pignut)** Pignuts grow mostly in woodlands and on sandy common land. The young foliage looks a bit like a delicate bracken. When I was a kid, we would search out the plant, and then lie down on the ground and use a knife or stick to dig down very gently alongside the stalk until we came to the nut-like root. The game was to see who could discover the biggest nut. Once you know what you are looking for, they are relatively easy to find. They are best eaten raw as part of a salad, or with bread and cheese.

Corylus avellana **(Hazelnut; Cobnut)** Grows in scrubby woodland and hedgerows to a height of about 13 ft. Best gathered in autumn when the nuts peep out from little, green, leaf-frilled husks. They are very tasty when chopped up and eaten with a tossed salad and brown bread and butter, or chopped and added to muesli, or chopped and added to a nut roast.

Lycoperdon **spp. (Puffball)** There are two types of puffball, both edible: the giant puffball (*Lycoperdon giganteum*) and the common puffball (*Lycoperdon perlatum*). Both have a white to cream, leathery skin and a spongy, creamy-white inner flesh. If it is bigger than about 4 in. in diameter then it is the giant puffball. If you find an example that is in any way yellow or brown in color, or scaly in texture, then it is either an old puffball or it is something else – either way it is not good to eat. Puffballs can grow to a dramatic size – as big as a football – so be ready for a feast.

Prunus spinosa (Sloe; Blackthorn)

Berry found in hedgerows that is small, round and dark purple-blue in color, and very good when made into sloe gin. If you spike the berries and cover them with sugar and gin, the resultant pink-gin liqueur is unforgettably tasty.

Pulmonaria officinalis (Lungwort)

A close cousin of comfrey, with spotted leaves and multi-colored flowers. Makes a tasty dish when steamed like greens and served up with brown bread and butter and a wedge of Stilton cheese.

Rorippa nasturtium-aquaticum (Watercress)

Watercress commonly grows in shallow, fresh, running water, in brooks and streams. The best way of identifying cress is to compare it with a commercially grown sample. Avoid the risks associated with wild cress – such as liver fluke – by only gathering it from woodland streams (not streams used by sheep and cattle) and by only picking and eating the small leaves and top shoots. I like it best in a sandwich; but, if you have worries about liver flukes, steam it and serve it as a side vegetable.

Rosa canina (Rosehip; Itching Berry)

Orange-red berry found in wild hedgerows that is very tasty when made into rosehip syrup. The skins are left to soften and then boiled with sugar and reduced until it turns syrupy. Very good for coughs and colds, it contains much more vitamin C than oranges.

Rubus fruticosus (Blackberry)

Grows in woods and hedgerows. Very prickly,

Wild fruits

with black berries. Just about everyone knows about this delicious, easy-to-find fruit. Perfect when made into jam.

Stellaria media (Chickweed; White Bird's Eye)

Small plant with diminutive, white, star-like flowers and minute green leaves that grows more or less anywhere. Instead of cursing at its abundance, you could try eating it. Very tasty steamed with chopped onions and olive oil.

Taraxacum officinale (Dandelion; Golden Sun; Clock Flower)

Although there are hundreds of traditional names for the common dandelion, a good number of them – bum flower, piss-phoo, mess-bed, wet-weed, shit-bed – seem to suggest that the dandelion was used in some way to control bowel and bladder movements. Whether they were used to make you stop or make you start is a mystery, but it is something you could experiment with. As a kid, I truly believed that if I picked a dandelion I would instantly wet my pants. Dandelions can be made into wine, a coffee substitute, and fried in butter as a micro vegetable all without risk to your pants.

Urtica dioica (Stinging Nettle; Devil's Leaf)

Of all wild plants, stinging nettles are one of the best recognized, the most common and the most painful to touch. From a very young age, most of us know and avoid nettles – we know how they sting, and how their touch results in an excruciatingly burning and bumpy red rash. Consequently, even though most people also know that nettles are good to eat, they cannot bring themselves to actually eat them. That is a pity, because they can be tasty, delicious even, in the form of soups, wine, cooked greens on the side, and in soups and teas. They are particularly nutritious and refreshing as an herbal tea.

75

SOIL CARE

The soil is a living, breathing entity that provides plants with shelter and sustenance. If it is any way sick or unbalanced – too acid, too alkaline, too cold, or in any way less than perfect – then the plants that look to it for food and shelter will suffer. If you stay with the ancient analogy of 'Mother Earth' then you will not go far wrong.

SOIL CHARACTER

When people talk about the 'character' of a soil – they might say that it is too heavy or sweet-smelling, or unforgiving, and such like – they are really talking about the sum total of its physical characteristics and the external forces that add or detract from those characteristics. So, while a soil might be sandy, clay, peaty, chalky or silty, it might also be wet, moist, waterlogged, dry and parched. For example, a waterlogged clay soil is more of a problem than a moist clay soil. While you cannot change the underlying physical characteristics – clay is clay – you can use those characteristics to best effect, and you can improve them.

SOIL LIFE

The soil is alive with all manner of living organisms – large items like worms and insects that we can see with the naked eye, and infinitesimally small organisms that we can only see under a microscope. It is not over-dramatic to say that every single seen or unseen life form in the soil does in some way rely on a neighboring life form for its well-being. If you destroy all the organisms that you perceive as being pests, you will also be destroying beneficial organisms.

ORGANIC MATTER

In life and in death, plants and animals are crucial to the well-being of the soil – in life they create organic matter, and in death they become organic matter. Without the ever-turning wheel that is life and death the soil would soon become sterile. Organic matter improves the physical structure of the soil, nourishes plants and other organisms in the soil, and absorbs and holds water.

SOIL MANAGEMENT

If we take it that soil can be abused, starved, neglected, over-used, and generally ill managed – and that such soil results in poor crops – then it follows that the task of the organic gardener is to care and nurture the soil by feeding it with organic material, covering it with a protective mulch, leaving it as far as possible undisturbed, and avoiding over-use.

COMPOST

A good compost heap demonstrates the eco-green, self-sufficiency, life-wheel philosophy at its best: garden and kitchen waste go in at one end, and soil improvers, plant foods and our food comes out at the other. One compost heap is essential, two are good, but three or more are a delight. A large, well-built, three-section compost unit is a great option (see above). Plenty of things can go in: weeds, garden waste, farmyard and stable manure, pet bedding, fruit and veggie scraps, paper and card, flower and grass clippings, natural fibers like wool and cotton – but not glass, tin cans, plastic, pet feces, dead animals, household chemicals, plants with diseases such as potato blight, uncooked meat, and pernicious weeds.

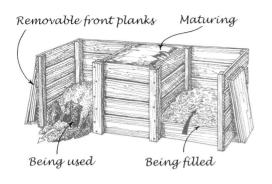

Removable front planks   *Maturing*

Being used   *Being filled*

Three-section composter

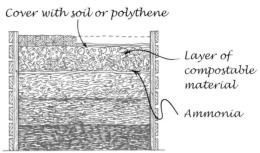

Cover with soil or polythene

Layer of compostable material

Ammonia

Compost layers

WORM COMPOSTING

Worm farms are fun, and will turn your kitchen scraps into liquid compost for your greenhouse plants and a friable compost to super-enrich your growing beds. You can either buy a ready-made worm farm, or make your own. Fit out an old plastic dustbin with a water-butt tap, add a little soil and several worm-rich shovelfuls of horse manure, and then start adding the kitchen scraps.

You will soon know if and when your wormery is working: it will smell good, the worms will multiply, and a turn of the tap will result in a very useful liquid feed. At the end of the process, you can empty out all the resultant high-quality compost and start the process all over again.

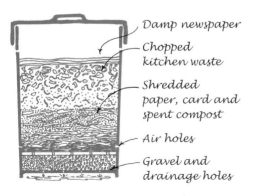

Damp newspaper

Chopped kitchen waste

Shredded paper, card and spent compost

Air holes

Gravel and drainage holes

Wormery composter

GREEN MANURES

Green manures are plants that are grown specifically to smother weeds, protect the soil with a living mulch, force out plant and animal pests by changing conditions, to fix nitrogen in the soil and be dug in to provide food for the soil and plants. For example, clover, mustard and trefoil sown in spring through to summer and dug in during early spring is a good option for giving body to a light to medium soil, and for increasing nitrogen levels.

ANIMAL MANURES

While farmyard animal manures from cows, goats, chickens and horses are the traditional option, the best manure for self-sufficiency is that which comes from organic farms or from your own self-sufficiency set-up. Animal manures – usually mixed with some sort of straw, paper or woodchip bedding – are a rich source of nutrients, perfect when composted with kitchen and garden waste. Cow manure is a wonderful source of all nutrients, poultry manures are rich in nitrogen but extra smelly, horse manure is rich in nitrogen and relatively easy to handle, pig manure is good but difficult to handle – each type of manure has its own unique quality. Animal manure can either be stacked and made into a manure heap or dug straight into the ground.

WATER FOR THE GARDEN

All roofs have potential for collection

Linked barrels with an overflow running to pond

Deep pond stores overflow and patio water

Water from greenhouse roof collects in butt

Trough

Patio drain to pond

Water collection and storage

Water is our most precious possession. In the context of self-sufficiency, everything starts with water storage, recycling and conservation. Working on the premise that using water from the utility company is both expensive and wasteful – it is not really logical to use high-grade drinking water to water plants – it follows that you need to cut out the expensive water-utility part by catching and storing the rainwater that falls on your plot before it runs off to the utility company.

RING CULTURE

This is a growing system that is designed to use water to best effect. The water-saving is achieved by the very simple process of sending the water directly to plants' roots. In essence, a ring-culture system consists of a plastic-lined trough, trench or bed full of a chemically inert aggregate – something like pea gravel and perlite, or granite chips and sand – with the plants being contained in

bottomless pots that are set directly on top of the aggregate. Nutrient-rich water is fed directly into the bed of aggregate, with the effect that roots pass out of the bottomless pot and run straight down to the underlying water. While water losses can be high in the early stages, this is offset by the fact that the whole system can easily be organized so that it uses saved rainwater. Some growers go one step further and minimize evaporation by covering the aggregate with a mulch of fine woodchip or paper covered in gravel.

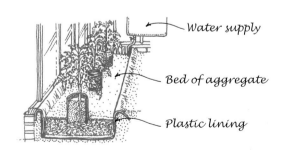

Water supply

Bed of aggregate

Plastic lining

Ring culture

SAVING RAINWATER

WATER BUTTS A water butt under each downpipe is a good traditional option; it is easy to set up and it is inexpensive. You could even use recycled butts. You could increase storage capacity by linking a number of butts. Set the butts up on concrete blocks so that the taps are high enough for a bucket or watering-can.

WATER TROUGHS One or more water troughs is a good option especially if you have livestock. The troughs can be decorative as well as functional.

UNDERGROUND CISTERNS These are good for retaining rainwater, but they are relatively tricky to build and expensive.

PONDS A pond is easy to build, it can be as big as money and space allows, and is decorative. Rainwater can be captured and diverted to a pond (see illustration opposite).

SWIMMING POOLS It is now possible to have a swimming pond or pool that uses an aerating system to keep the water fresh. You could build such system into your scheme, and pump the water around the garden.

USING WATER TO BEST EFFECT

MULCH Cover the surface with mulch – bark, leafmould, straw and so on – to hold in the moisture and to decrease the rate of evaporation.

ORGANIC MATTER Dig in bulky organic matter so as to increase the soil's water-holding capacity; this could be straw, manure, well-rotted bark, or anything that will hold water and eventually rot down.

RING CULTURE Using open ring pots in conjunction with a trough (see left, below) is a good way of directing water straight to the roots; it cuts down on loss by evaporation.

HOE MULCHING Hoe mulching is a traditional technique that involves repeatedly stirring the top 1 in. of the soil with a hoe in dry weather; the idea is that the broken-up surface reduces the soil's ability to give off water.

SHELTER Use wind brakes to cut down on the evaporation caused by drying winds, and awning to keep off the sun.

DROUGHT-RESISTANT VARIETIES Plant varieties that flourish in dry conditions.

SELECTIVE WATERING Do not bother to water lawns; it is a waste of water.

WATERING TIMES Water first thing in the morning or last thing in the evening. Watering in the evening will attract the slugs, but you can either kill the slugs or plant slug-resistant varieties.

SOAKER HOSES Hoses pierced by lots of holes or slots and buried just below the ground are not only a good way of ensuring that the water is directed straight to the target, but they also help to cut down on evaporation, a bit like ring culture.

GRAY WATER You can use gray water from showers and the kitchen sink, as long as it is free from too much soap and detergent. It cannot be used, however, if it contains oil, bleach, color or chemicals. Gray water soon goes bad, so you must remember to use it as soon as possible.

WEED CONTROL

It does not really matter how you look at it, or how organic you are, or how experienced you are as a gardener – a weed is a weed. You may not know its name, but if one plant is growing where you want to put another, then it is by definition a weed.

DEALING WITH WEEDS

In the context of self-sufficiency and organic gardening you have a choice: you can eat them if they are edible (for example, nettles, blackberries, dandelions and chickweed can all be eaten; see pages 74–75), let your animals eat them, dig them up, hoe them off, cut them off, starve them out, cover them up and plant over them, crowd them out, burn them off, and last but not least change your attitude towards them and consider them in a new positive light – because they are pretty, or because they attract beneficial insects.

HOEING

There is something very positive, pleasant and straightforward about hoeing – chop-drag, chop-drag. All you do is stir the top 1 in or so of soil with the hoe of your choice, and the job is done. Traditionally, there are beautifully

shaped, swan-necked draw hoes that you work with a delicate, dragging action, Dutch hoes that you work with a little push-and-pull action, large hoes with adze-like heads, large hoes like a cross between a hoe and a hatchet, and so on. Start with something like a good-quality swan-necked hoe, and then move on to other hoes when you have a better understanding of your needs. You will soon get to know, for example, that a well-shaped swan-necked hoe is the perfect tool for working between rows, while a Dutch hoe is good when you want to slice weeds off just below the surface of the soil.

RECYCLED MULCHES

Recycled mulches can be just about anything from sheets of paper and cardboard through to wood chips, sawdust and shredded paper. If it can be used to cover the ground (either below or above the surface), organic and will eventually break down, it can be defined as recycled mulch. For example, we use a thick layer of shredded bark to cover the paths that run between the raised beds in our vegetable plot. The paths are comfortable and safe to walk on, the mulch holds back the weeds, and

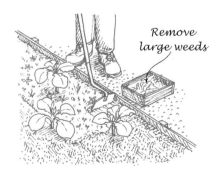

Remove large weeds

Hoeing

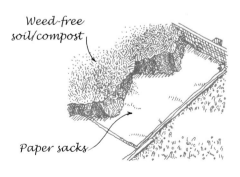

Weed-free soil/compost

Paper sacks

Recycled paper sheet mulch

best of all, when the bark has broken down after a year or so, we simply shovel it onto the beds, dig it in and put down fresh bark.

LOOSE MULCHES

Loose mulches might be anything from sawdust or minced paper through to bark, straw, hay, crushed shells and pebbles. Much depends on the context, and whether you are going to spread the mulch over a plastic sheet membrane, but in the food garden I always choose a mulch that will eventually break down. Pea gravel on a plastic membrane works, but the gravel will eventually gravitate off the membrane.

INTERCROPPING

While intercropping is generally thought of as a growing system that is used to maximize the growing potential of a plot – for example, you might sow fast-growing radishes alongside slow-growing celery – it can also be used in much the same way as living mulch. It is pretty straightforward: if you spot a space, simply plant fast-growing crops in it so that weeds are crowded out.

USING A GREEN MULCH

Green mulches are an excellent option in that they not only crowd out weeds, but also enrich the soil for the next crop. Say you have a patch of ground – a vegetable bed – that has to be left unplanted for a few months, and you do not want the weeds to take over.

1 Choose a fast-growing green manure, something like alfalfa, clover or field beans.
2 Sow your chosen seed closely, so as to cover the ground completely.
3 When the plants have reached the young leaf stage, chop them up and dig them into the first spit of soil.

USING A MEMBRANE

If you have beds that are chest-high in weeds, and you want to get started fast, without any digging, try using a woven plastic membrane.

1 Take a scythe and chop all the weeds down.
2 Take a spade and systematically go over the plot further, stomping and chopping, until you have an even mulch of chopped weeds.
3 Slice up a lot of cardboard packing cases and unwanted paper, and spread them over the chopped weeds.
4 Top the whole lot with a woven plastic membrane and fix it down at the edges with pegs, boards, concrete blocks or whatever comes to hand.
5 Finally, cut some little crosses through the membrane and the card/paper, and sow or plant your chosen crop through the holes.

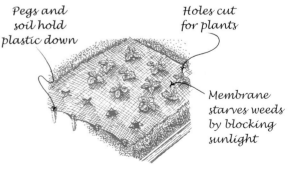

Mulch of spent manure

Moisture retained for roots

Loose mulch

Pegs and soil hold plastic down

Holes cut for plants

Membrane starves weeds by blocking sunlight

Woven plastic membrane

CROP ROTATION

One of the first rules of agriculture is that the same kind of crops should not follow, one after another, on the same kind of soil. So, for example, you should not keep planting potatoes in the same plot. If you do, the crop will keep using the same food elements until the soil becomes depleted and the plants become sick and diseased.

A GOOD LAYOUT

Divide your land into four plots:

- **Plot 0** – Permanent crops
- **Plot 1** – Brassicas
- **Plot 2** – Legumes and salad crops
- **Plot 3** – Root vegetables

Set the land out so that the rows within the individual plots run in a north–south direction, with the permanent plot on the windward side. Keep a plan or diary of the cropping, so that you can refer to past years when you are making plans for the years to come.

THREE-YEAR ROTATION PLAN

Divide the ground up into four plots numbered 0, 1, 2 and 3. Put plot 0 aside for permanent crops and specialist vegetables, and manage the other three plots on a three-year rotation. This scheme can be taken one step further by rotating the crops within the individual plots. For example, even though the roots will be grown in plot 1 the first year, plot 2 in the second year, plot 3 in the third year, and back to plot 1 in the fourth year, when you come around to planting the roots in plot 1 you can plant them in a different position, so that say carrots follow potatoes, rutabagas follow beetroots, and so on.

PLOT 0 – PERMANENT VEGETABLES AND HERBS

The following crops are defined as 'permanent' in the sense that they can remain in or near the same bed for a number of years.

GLOBE ARTICHOKE reaches its peak in the third or fourth year.

ASPARAGUS can be left in the same bed for 10–20 years.

BAY a hardy evergreen shrub that likes a well-drained, moisture-retentive soil in a sunny position.

BORAGE a hardy annual.

CHERVIL a hardy biennial that is usually grows as an annual.

CHIVES a hardy, low-growing, clump-forming perennial.

DILL a hardy annual.

FENNEL a hardy herbaceous perennial.

MINT a hardy herbaceous perennial.

PARSLEY a hardy biennial.

RHUBARB can be left in the same bed indefinitely.

ROSEMARY an evergreen shrub.

SAGE a hardy evergreen shrub.

THYME a hardy dwarf evergreen shrub.

SPECIALIST VEGETABLES The following crops are defined as 'specialist' for no other reason than that they can be grown in the permanent plot:

- **Eggplants**
- **Capsicums (sweet peppers)**
- **Marrows and courgettes (zucchini)**
- **Tomatoes**

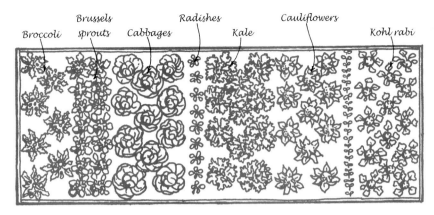

Broccoli · Brussels sprouts · Cabbages · Radishes · Kale · Cauliflowers · Kohl rabi

PLOT 1, 1ST YEAR, BRASSICAS and plants that enjoy
the same soil (plot 2 in 2nd year; plot 3 in 3rd year)

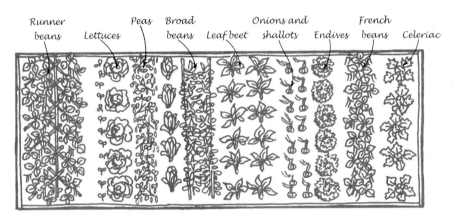

Runner beans · Lettuces · Peas · Broad beans · Leaf beet · Onions and shallots · Endives · French beans · Celeriac

PLOT 2, 1ST YEAR, LEGUMES AND SALAD CROPS
(plot 3 in 2nd year; plot 1 in 3rd year)

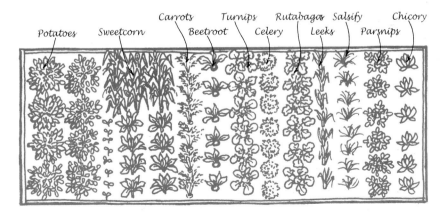

Potatoes · Sweetcorn · Carrots · Beetroot · Turnips · Celery · Rutabagas · Leeks · Salsify · Parsnips · Chicory

PLOT 3, 1ST YEAR, ROOT VEGETABLES and plants that enjoy
the same conditions (plot 1 in 2nd year; plot 2 in 3rd year)

DIGGING

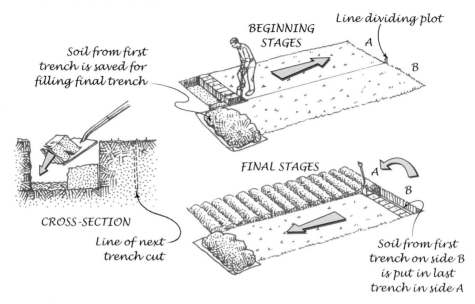

Soil from first trench is saved for filling final trench

BEGINNING STAGES

Line dividing plot

A
B

CROSS-SECTION

Line of next trench cut

FINAL STAGES

A
B

Soil from first trench on side B is put in last trench in side A

Traditional single digging

Traditionally, the ground was dug in autumn and spring, the idea being that the cutting and slicing action of the tools, combined with the weathering action of the sun, wind, frost and rain, variously breaks down the organic matter in the soil, kills pests, aerates the soil and increases drainage. Current thinking is that, while digging is undeniably beneficial in the short term, in the long term it damages drainage, compacts the soil, breaks down the positive structure of the soil and generally decreases fertility.

TRADITIONAL SINGLE DIGGING

Divide the plot down the middle of its length. Take out one spit or trench – the depth and width of the spade – from the first half and lay it down to the side. Add manure or compost to the trench bottom. Remove the thin layer of weeds from the second trench and lay it upside down in the first trench. Take out the soil from the second trench to the depth of one spit and lay it over the weeds in the first trench. Repeat the procedure down to the end of the plot. Move to the second half, take out one spit, and continue in the manner just described. When you get to the end of the second half, put the contents of the first spit to be dug into the last trench.

ORGANIC NO-DIG RAISED-BED CULTIVATION

With digging being a traditional method of using hand tools to physically turn over the soil in preparation for planting, the no-dig raised-bed method is an organic technique of preparing the soil for planting that involves leaving the base soil undisturbed. In many ways, it is the complete opposite of everything that traditionalists advocate as being positive. The benefits of the no-dig raised-bed method

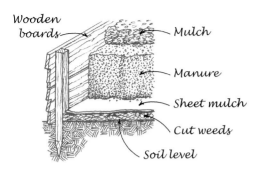

No-dig method

include not so much sweat and toil, the soil never gets disturbed, and, best of all, fewer weeds are brought to the surface.

NO-DIG METHOD

1 Use wooden boards to build frames for raised beds about 4 ft square and 12–18 in. high.

2 Cut down the weeds inside the frames and cover the soil with a sheet mulch of paper/card topped with well-rotted manure.

3 Top it all with a layer of loose mulch, such as straw, hay or wood chips.

TRADITIONAL DOUBLE DIGGING

Double digging aerates the deeper layers of the soil, resulting in bigger roots and more vigorous plants. Divide the plot down the middle of its length. Take out two spits so as to finish up with a trench about 2 ft wide, and put it on one side. Add manure or compost to the trench bottom and turn it over to the full depth of a fork. Skim the turf about 2 in. deep from the next two spits and lay it upside down in the first trench. Take out the soil from the next two spits and lay it over the turf so as to fill up the first trench. Repeat the procedure (see below) until you get to the end; then put the contents of the first two spits into the last trench.

PREPARING A TRENCH

For beans or marrows, dig out an 18 in wide trench one spit deep, and put the soil to one side. Fork well-rotted manure and/or compost into the bottom, to the full depth of the fork. Return the topsoil to the trench.

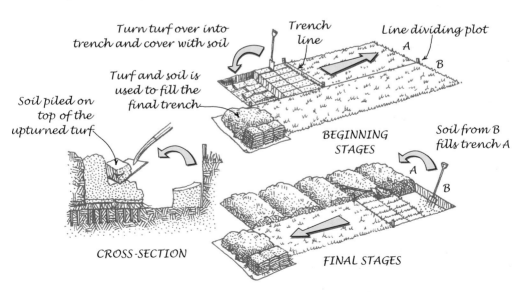

Traditional double digging

TRACTORS, TILLERS AND ROTOVATORS

If you have anything more than an acre, the time will almost certainly come around when you will want to cultivate a piece of land. You could opt for a tiller or a rotovator (and they are fine for a small plot) or a garden mini-type tractor – but these are very expensive and not really up to the task. So what to get? The cheapest and most efficient solution for most small set-ups is to get a 1950s-type tractor. These tractors are winners on many counts – they were built to last a lifetime, there are lots of them around, they are made from heavy cast iron, they have a slow running speed, their basic design allows for easy maintenance, and best of all they are low-priced. A small secondhand 1950s-type tractor – a Ferguson, John Deere or Ford – can be used to plough and harrow, cut grass and hedges, bang in posts, lift and carry loads, and the list goes on.

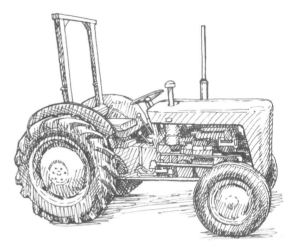

Ferguson tractor

BUYING A SECONDHAND TRACTOR

Talk to as many old farmers as possible, tell them about your acreage, explain what you want the tractor to do, and generally ask for advice. In the light of your research, and having decided on the make and model, get an owner's manual. Having had a long slow read through the manual, go and see what is on offer. Always explain that you want to start the tractor from cold. When you have the tractor in front of you, work through the following procedures and checklist.

- Is it a pre-60s tractor – one of the big names like Ferguson, Ford or John Deere?
- Start the tractor. Does it start easily? It is vital that you start it from cold. Be wary if it has been warmed up. If it does start easily, then at least you know that items like the battery, ignition system, magneto and fuel pipes are in good order.
- With the engine still running, have a look at the exhaust smoke. Is the smoke black, white or blue? Don't worry too much if it is black or white, but do worry if it is blue – this suggests that items like rings and valves might be less than perfect.
- With the engine still running, spend time looking for oil, fuel and water leaks.
- Listen to the engine. If there is a heavy clunking or clattering noise, the chances are that the engine will need to be stripped down and items like bearings, rods and crankshaft will need to be overhauled.
- Run the engine for about an hour, and then shut it down and try to restart.

- Stop the engine, wait until it is cool, and then check the oil and water. If the water is black and scummy, or the oil white and foamy, look for another machine.
- Look the whole machine over for cracks. Don't worry about 'soft' items like the seat, or the pressed steel bits, or the paintwork, just be on the lookout for cracks in the heavy cast items.
- Tires are expensive, so give them a good checking over. Look for cuts and cracks that may indicate that the tractor has had some rough treatment.
- Rollover bars are an essential safety feature. If your chosen tractor has not been equipped with such devices – and the chances are a 1950s tractor will not have them – then be sure to have them fitted. You don't have to have them fitted if you and the family are sole users, but it is a good idea anyway – especially if you are a raw beginner.

FUEL OPTIONS

Depending on the make, model and age, old tractors are designed to run on petrol, diesel or kerosene. Each type of fuel has its champions. Some say that kerosene is the most economical; others say that petrol is more reliable, for example. I prefer diesel because of its low cost and availability.

TRACTOR IMPLEMENTS

Having purchased your tractor, and having been shown by an expert how to generally stop, start and run the machine, then comes the exciting business of kitting it out with the various implements. Apart from the plough and disc, there are grass scythes, buckets, balers, wood saws and so on. For example, with our little acre plot, all we really needed

was the box or bucket and the scythe, both of which were fitted to the three-point linkage.

TILLERS, ROTOVATORS AND SIT-ON CULTIVATORS

These machines come in all sizes and shapes – some good, some not so good and others just plain horrible. The first thing you have to remember here is that really they are only good for cultivating the top 6 in. or so of the soil. My experience suggests that they are hard work, bad on the back, always going wrong, expensive and short-lived. Of course, some machines are better than others, but really the only way to judge if they are going to suit your needs is to visit a supplier and have a trial run.

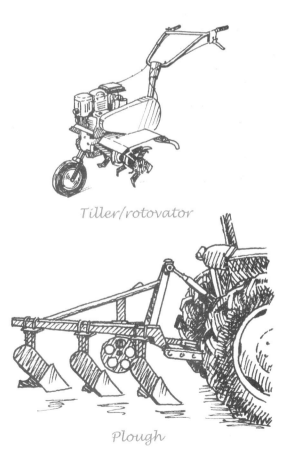

Tiller/rotovator

Plough

SOWING AND PLANTING

With some vegetable crops, it is best to sow seeds indoors and nurture the seedlings before planting them out in the open garden. With other crops, such as carrots and radishes, you would normally sow the seeds in drills and then thin out the seedlings. Other seeds, such as those of cabbages, are sown and then either thinned out and left in place or transplanted as seedlings to a final growing position. Seeds of peas and beans can be sown at a chosen spacing and left alone. Others might be sown in containers, and later transplanted to the open garden. To a great extent, the chosen method depends on the size of the seeds and the innate character of the plants being sown (see pages 116–143 for advice regarding individual crops).

3 Use a line to mark in the planting position, and with a swan-necked draw hoe carefully draw out a shallow drill, at a depth to suit your chosen seeds.

4 Moisten the soil with a watering-can fitted with a rose, and use your fingertips to dribble seed along the drill.

5 Lightly cover the seeds with soil and use a rake to tamp it level.

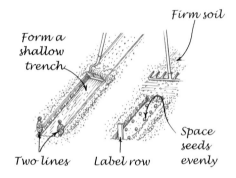

Firm soil

Form a shallow trench

Two lines *Label row* *Space seeds evenly*

Medium seeds

SOWING MEDIUM SEEDS IN OPEN GROUND

1 Use a line to mark in the planting position, and with a swan-necked draw hoe carefully cut a shallow trench or trough, at a depth to suit your chosen seeds.

2 Moisten the soil with a watering-can fitted with a rose, and carefully place each seed in position.

3 Rake a layer of soil over the seeds and tamp it level.

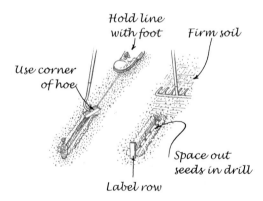

Hold line with foot *Firm soil*

Use corner of hoe

Space out seeds in drill

Label row

Small seeds

SOWING SMALL SEEDS IN OPEN GROUND

1 Break the soil down into a fine tilth – meaning a fine, crumbly texture.

2 Work backwards and forwards with the rake, removing any stones, bits of twig, hard lumps of earth, and so on.

SOWING LARGE SEEDS IN OPEN GROUND

1 Break the soil down into a fine tilth – work backwards and forwards with the rake, removing any stones.

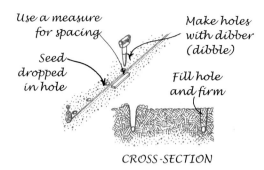

Use a measure for spacing

Make holes with dibber (dibble)

Seed dropped in hole

Fill hole and firm

CROSS-SECTION

Large seeds

2 Use a line to mark in the planting position, take the rake and tamp the ground firm.

3 Moisten the soil with a watering-can fitted with a rose, and use a dibber to plunge holes; the size of the dibber and the depth of the holes should suit your chosen seeds.

4 Drop the seeds into the holes, water them in, lightly cover with soil and use a rake to tamp level.

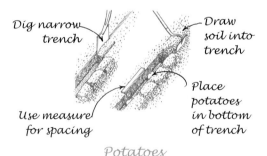

Dig narrow trench

Draw soil into trench

Place potatoes in bottom of trench

Use measure for spacing

Potatoes

PLANTING SEED POTATOES IN OPEN GROUND

1 Break the soil down into a medium tilth, removing large stones.

2 Use a line to mark in the planting position, and use a small spade to cut a trench to a depth of 4–6 in.

3 Moisten the soil and carefully set each potato in place.

4 Cover the potatoes with soil and use a hoe to earth up.

SOWING SEEDS IN CELLS

This method minimizes handling damage.

1 Fill the cellular block or tray (flat) with a compost/soil mix to suit your seeds.

2 Use a batten to skim away excess compost and firm up the contents of each cell so that the level is just below the top of the cell.

3 Moisten the compost, sow 1–2 seeds in each cell, and lightly cover.

4 When the seedlings are well established, ease them out complete with the block of peat and plant them out.

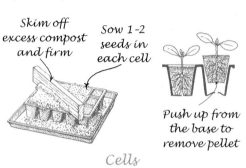

Skim off excess compost and firm

Sow 1-2 seeds in each cell

Push up from the base to remove pellet

Cells

SOWING SEEDS IN PEAT POTS

This method minimizes root disturbance.

1 Fill the peat pot with a compost/soil mix to suit your chosen seeds.

2 Moisten the compost, dib a hole, and set 1–2 seeds in place.

3 Cover with compost and lightly water.

4 Thin the seedlings to the best plant.

5 When well established, plant the whole pot and the young plant in its final position.

Prick out seedlings into peat pots

Plant out complete with pot

Peat pots

GROWING UNDER COVER

In a greenhouse or polytunnel you can garden whatever the weather. Other advantages are that you can have a year-round season for a wide selection of crops, you can grow new or unusual crops, and you can create a micro-environment to suit the needs of your crops. Crops can also be sown earlier and harvested later than those planted in open ground.

GREENHOUSES

A greenhouse is the traditional option for a small garden – it is relatively low in cost, there are lots of designs to choose from, glass is a

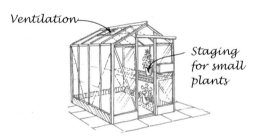

Ventilation

Staging for small plants

Greenhouse

relatively easy material to source and replace, and there are all manner of DIY kit options. A greenhouse can be a great asset, the proviso being that it is made of cedar and at least 8 ft wide.

MANAGING A GREENHOUSE

- Keep the greenhouse clean by tidying up as soon as you make a mess.
- Make sure when you are picking and trimming shoots and leaves (as with, say, tomatoes) that you remove the trimmings from the greenhouse.
- Disinfect the inside at the start of the season, especially in corners and under benching; use an eco-friendly disinfectant.

- Water in the morning, rather than in the evening.
- In very hot weather, soak the paths and surfaces in the greenhouse so as to achieve a cool, 'damped-down' humidity.
- On bright days, shade by spreading screens or close-mesh nets on the sunny side.
- In the winter, insulate by fixing bubble wrap over the inside. This is much easier to do with a wooden greenhouse.
- Make sure that there is plenty of ventilation, especially in hot, humid weather.

POLYTUNNELS

Polytunnels are a relatively new phenomenon – a good choice when you want, as it were, to do your outdoor gardening indoors. The average polytunnel consists of a series of bridge-shaped, galvanized steel hoops covered with clear plastic sheet.

Although there are lots of options to choose from (high, long, low), a short, wide configuration, with high, straight sides and a wide door at each end is a good choice, in that it gives the best possible headroom, growing space, access and ventilation, as well as allowing for a central walkway.

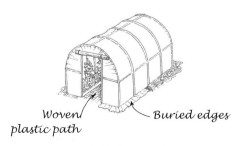

Woven plastic path

Buried edges

Polytunnel

MANAGING A POLYTUNNEL

- In warm weather, roll sheets of plastic over the end doors to lock in the heat.
- In breezy weather, close the screen down on the windward ends to prevent cold air 'tunnelling' through.
- Lift foliage away from the inside faces of the tunnel to prevent it dying off and eventually rotting.
- Be careful when using canes, wire and similar sharp items that you do not cut through the plastic.
- Be careful if you are using a high-pressure water spray that you do not blow a hole through the plastic sheet.

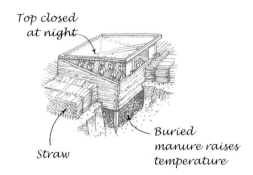

Top closed at night

Straw

Buried manure raises temperature

Hotbed

COLD FRAMES AND HOTBEDS

Cold frames and hotbeds – both types of glass-covered frame – are the traditional answer to growing plants under cover. A hotbed is a cold frame that has been super-insulated, traditionally with straw and rush matting all around, and placed on a pit or bed full of fresh farmyard manure. The manure breaks down slowly and releases heat, with the effect that the micro-climate within the frame becomes hot. Hotbeds are now usually covered with materials such as old carpet and/or plastic bubble wrap.

MANAGING COLD FRAMES AND HOTBEDS

- If you are short of money cold frames can easily be built from salvaged wood and plastic sheet.
- They are easy to ventilate.
- In times past 'French beds' were considered to be a revolutionary system of growing produce out of season. The idea was to have line upon line of hotbeds. This is a wonderful system to try if you have unlimited supplies of fresh horse manure.
- For hotbeds, it is essential to use fresh manure and to bury it as soon as possible after collection as it soon loses its heat if left uncovered.

GROW HOLES

A grow hole was traditionally the poor man's answer to a hotbed. At its simplest, it consists of a deep pit with a thick layer of fresh farmyard manure in the bottom, a collection of stout cut saplings long enough to span the pit, and one or more woven rush mats. The manure breaks down and produces heat, just like the hotbed. At night, and/or if the weather turns cold, the saplings are placed over the pit, and the whole thing is covered with the matting.

MANAGING GROW HOLES

- Grow holes can be difficult if the ground is sodden or the water table is high. If this is the case, maximize drainage by digging the hole on high ground and including a sump hole under the manure layer.
- Grow holes used with square straw bales are good, in that you can build protective walls on the windward side.
- If the soil is soft and sandy, shore up the vertical faces of the hole with metal sheet.

GROWING METHODS

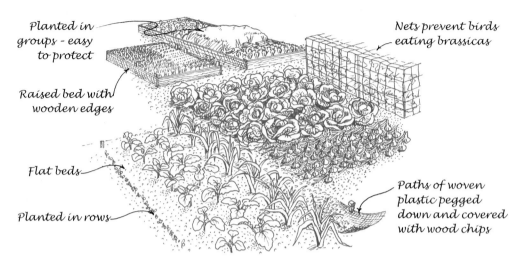

Planted in groups - easy to protect

Raised bed with wooden edges

Flat beds

Planted in rows

Nets prevent birds eating brassicas

Paths of woven plastic pegged down and covered with wood chips

Growing methods

Experts now consider that the traditional system of growing vegetables in straight lines or rows, with a space for a path between rows as well as a generous space between plants, is faulty. The problem is that, once the plot has been dug, planted, tended and harvested, the initial cost and energy that has been given to the ground between the rows will not only have been wasted, but along the way the structure of the soil will also have been destroyed by compaction.

Put another way, half of the money, effort, time, ground and even the manure will have been wasted. The various bed systems described below have been designed to improve upon traditional practices.

FLAT BEDS: ADVANTAGES AND DISADVANTAGES

Flat beds are raked and levelled areas.
- Once the beds are established, you will not be wasting time and energy on digging and improving the paths.

- The set-up costs are low but the gradual build-up of soil will spill onto the paths.
- If you aim for a bed width of about 4 ft, you will be able to work without treading on the bed.
- The close spacing cuts down on the number of weeds.

RAISED BEDS: ADVANTAGES AND DISADVANTAGES

Raised beds are containments that raise the growing area above ground level.
- Once the beds are established, with narrow paths all around them, you will be able to concentrate on production.
- The set-up building costs are high, but all high-quality soil will be contained.
- You will not need to tread on the beds.
- The close spacing of the planting cuts down on the number of weeds.
- You can build up the height of beds and the depth of the soil to suit your needs.
- You will not have to do any heavy digging.

SQUARES: ADVANTAGES AND DISADVANTAGES

Squares allow plants to be grown in a mass rather than in rows.

- Once the beds are established, with narrow paths all around them, you will be able to concentrate on crop production.
- The set-up costs of building the raised edges are high, but the high-quality soil will all be contained.
- All the work can be done from the paths.
- Each bed can be reserved for growing a specific crop.
- Crop rotation will be simplified.
- The number of weeds will be restricted to the minimum.
- You will more easily be able to protect the beds, with net, fleece or twigs.
- You can build up the height of individual beds and the depth of the enclosed soil to suit your various needs.
- The shape and pattern of the beds will become a design feature in their own right.

TRADITIONAL SOIL PREPARATION

SPRING – CONSOLIDATING

On a crisp dry day in early spring, when the ground is dry underfoot, take the rake and work systematically over the ground breaking up clumps of soil. Tamp the resultant tilth to an even finish. If the soil sticks to your boots or the rake, it is too wet.

SPRING – LEVELLING

Work backwards and forwards over the ground, skimming off high spots, filling in hollows and removing debris. Do not attempt to remove every last little pebble, because they are good for drainage; just take out the big ones.

AUTUMN – FORKING AND WEEDING

On a crisp day, when the ground feels dry underfoot, take a fork and lightly work the ground. Remove weeds and debris.

AUTUMN – IMPROVING WITH MANURE

Spread a generous mulch of manure over the ground – the more the better.

PLANNING THE YEAR

This calendar will not answer all your queries but it will point you in the right direction.

MID-WINTER

GENERAL TASKS Inspect your tools, stakes, pots, plastic sheets and fleece. Set seed potatoes to sprout. Order seeds. Plan the plots for sowing. Prune hardy orchard trees.

SOIL PREPARATION Turn over vacant soil or spread a mulch to cover. Spread manure.

SOWING AND PLANTING Plant broad beans in a warm spot. Sow onions, leeks, radishes in a frame or hotbed.

HARVESTING AND STORING Pick sprouts, winter cabbages, last carrots, celery, chicory, and anything else that is ready. Bottle or freeze carrots. Bottle celery. Salt cabbages.

LATE WINTER

GENERAL TASKS Clean up paths. Cover frames at night.

SOIL PREPARATION See if you can get ahead with more digging or mulching. Prepare seed beds. Look for a warm corner and make sure there is a good even tilth. Remove weeds. Put rubbish in the compost heap.

SOWING AND PLANTING Plant artichokes and shallots. Sow early peas and another row or two of broad beans. Sow carrots, lettuces and radishes under glass on a hotbed. Raise seedlings in a warm frame.

HARVESTING AND STORING Pick Brussels sprouts, winter cabbages, last of the carrots, celery, chicory, and anything else that is ready. Freeze or bottle carrots. Bottle celery. Salt cabbages.

EARLY SPRING

GENERAL TASKS Mend fences. Look at your plot and see if you want to change things around.

SOIL PREPARATION Break down and rake the surface of double-dug plots. Keep stirring with the hoe.

SOWING AND PLANTING Sow hardy seeds like lettuces and parsnips out of doors. Sow things such as spinach, broccoli, leeks, onions, peas, celery, tomatoes and marrows under glass, either directly in the bed or in trays (flats).

HARVESTING Pick Brussels sprouts, cabbages and cauliflowers. Freeze or bottle cauliflowers. Salt cabbages.

MID-SPRING

GENERAL TASKS Check for slugs and snails. Watch out for problems on fruit trees and bushes. Thin seedlings as necessary and keep weeding. Cover cold frames at night.

SOIL PREPARATION Keep working the ground with the hoe, especially alongside seedlings. Draw soil up on potatoes.

SOWING AND PLANTING Sow just about anything in the open. Plant maincrop potatoes, onions, radishes, maincrop carrots, beet, salsify and scorzonera, endives, kohl rabi, lettuces, peas, and spinach. Plant out any hardened-off seedlings. Sow runner beans, marrows, zucchini under glass.

HARVESTING AND STORING Pick beet leaf and broccoli. Freeze or bottle broccoli.

LATE SPRING

GENERAL TASKS Protect tender seedlings. Watch out for blackfly on beans. Set twigs among the peas. Put mulch around fruit trees and bushes. Reduce runners on strawberries. Water seedlings. Keep hoeing and weeding.

SOIL PREPARATION Prepare more seed beds. Hoe and rake regularly. Earth up potatoes. Mulch between rows of advanced vegetables.

SOWING AND PLANTING Plant out hardy seedlings from the frames. Sow tender vegetables in the open. Sow French, runners and broad beans. Sow more peas, endives, radishes and summer spinach. Plant out Brussels sprouts, broccoli and cucumbers.

HARVESTING AND STORING Pick beet leaf, broccoli, early beetroot, early carrots, cucumbers under glass, endives and numerous other vegetables. Freeze or bottle broccoli and carrots. Bottle beetroot. Salt cabbages.

EARLY SUMMER

GENERAL TASKS Bring in fresh manure. Keep watering. Spread mulches around turnips. put nets over fruit. Remove weak canes from the raspberries. Clean out cold frames. Keep hoeing and weeding. Stake runner beans and peas.

SOIL PREPARATION Keep hoeing. Dig up potatoes. Fork over vacant seed beds.

SOWING AND PLANTING Plant out seedlings from the nursery beds. Sow succession crops such as endives, lettuces and radishes.

HARVESTING AND STORING Pick anything that takes your fancy, such as salads, summer spinach, peas and early potatoes. Freeze, salt and bottle as much as possible.

MID-SUMMER

GENERAL TASKS Stake plants that look weary. Gather soft fruits when ready. Cut mint and herbs for drying. Topdress with manure. Look at the tomatoes and pinch out and feed as necessary. Lift and dry potatoes. Keep hoeing between crops. Water and weed. Make sure that the greenhouse and frames are open to the air.

SOIL PREPARATION Weed and hoe. Weed after lifting potatoes. Earth up maincrop potatoes.

SOWING AND PLANTING Plant out celery, cabbages, Brussels sprouts and broccoli.

HARVESTING AND STORING Keep picking, eating and storing. Freeze or bottle broccoli. Salt cabbages. Bottle beetroot. Dry or bottle shallots. Bottle a few choice potatoes.

LATE SUMMER

GENERAL TASKS Order seeds for autumn sowing. Keep bottling, drying and freezing vegetables and fruit. Bend over the necks of onions. Dry herbs. Pinch out the tops of tomatoes. Clear and dig the ground. Protect fruit crops from the birds. Plant out new strawberry beds. Keep forking, hoeing, weeding.

SOIL PREPARATION Weed and hoe. Dig and fork over empty potato plots.

WEEDING AND PLANTING Sow endives, radishes, spinach, onions and any other seasonal crops. Sow lettuces and salad crops under glass. Sow cabbages for spring planting.

HARVESTING AND STORING Keep picking, eating and storing. Dry more herbs. Gather beans, tomatoes and fruit as and when they are ready. Store roots in sand. Freeze, bottle or salt broad beans. Gather early apple. Salt cabbages.

EARLY AUTUMN

GENERAL TASKS Protect crops from frosts if necessary. Lift and store roots. Earth up celery and leeks. Watch out for and destroy caterpillars. Prune raspberries. Water ,weed and hoe as necessary. Blanch endives.

SOIL PREPARATION Weed and hoe. Dig, fork, and rake when you have cleared the crops. Earth up plants as needed.

SOWING AND PLANTING Plant out spring cabbages. Look at your seed packets and sow if possible.

HARVESTING AND STORING Lift potatoes and onions. Gather runner beans. Lift and store roots. Gather and store tree fruits as they ripen. Keep picking and eating other crops as and when ready. Salt or freeze runner beans. Dry or pickle onions. Lift and store roots. Pickle green/ripe tomatoes.

MID-AUTUMN

GENERAL TASKS Watch out for frost and protect as needed. Start digging vacant plots. Continue hoeing and weeding. Clear the ground and add to the compost heap. Clean up paths and edges. Thin onions.

SOIL PREPARATION Weed and hoe. Dig and fork over plots as they become vacant.

SOWING AND PLANTING Plant rhubarb and fruit trees. Sow peas in a protected cold frame. Sow salad crops under glass. Plant out seedlings. Sow early peas in warm areas.

HARVESTING AND STORING Gather the rest of the tomatoes. Lift and pick celeriac and carrots. Pickle tomatoes and beetroot. Gather and store apples and pears.

LATE AUTUMN

GENERAL TASKS Watch out for frost and protect as needed. Clean up leaves and debris. Dig vacant plots. Hoe and weed as necessary. Remove bean and pea sticks and poles.

SOIL PREPARATION Weed and hoe. Dig and fork over plots as they become vacant.

SOWING AND PLANTING Sow broad beans in a sheltered spot.

HARVESTING AND STORING Lift and store root crops in sand. Cut, lift and eat other crops as needed.

EARLY WINTER

GENERAL TASKS Watch out for frosts and protect as needed. Clean the tools and the shed.

SOIL PREPARATION Weed and hoe. Dig and fork over plots as they become vacant. Fork the soil over so as to expose pests. Check that stored vegetables are in good order.

SOWING AND PLANTING Plant broad beans. Draw earth up around the peas. Sow salad crops under glass and protect if needed.

HARVESTING AND STORING Pick the last of the beet leaf. Pick Brussels sprouts, winter cabbages, last of the carrots, celery, chicory and anything else that is ready.

POTATOES

Potatoes are arguably the most popular vegetable. They are relatively easy to grow and to store, they are nutritious (very high in vitamin C and potassium), they can be cooked in many different ways (roasted, mashed, chipped, baked in their skins, boiled, sliced and fried) and they are altogether tasty. Of course, fresh homegrown 'spuds' – new potatoes with mint sauce or old potatoes roasted – are many times tastier and more nutritious than shop-bought ones. In the UK and the USA the average adult eats about 130 lb of spuds a year. So a family of four would need 520 lb of potatoes. The good news is that most kids like potatoes. In the context of self-sufficiency, potatoes are an important food crop. The different varieties are grouped according to optimum planting time, so
there are 'first earlies,' 'second earlies' and 'maincrop' potatoes.

CHITTING

Chitting is a method of forcing seed potatoes to sprout prior to planting out. In late winter,

take your organically grown, disease-resistant seed potatoes and arrange them on end in shallow boxes so that the end showing the most 'eyes' – the 'rose' end – is uppermost. Store the boxes in a light, airy, frost-free room for about 5–6 weeks. When the potatoes show lots of shoots, rub the weak ones off to leave the strongest 2–3 sprouts per potato.

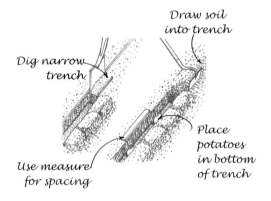

Planting potatoes

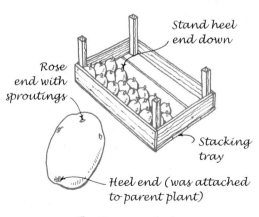

Chitting potatoes

TRADITIONAL PLANTING

Dig a shallow trench, or make holes about 1 ft apart and about 2–3 in. deep. Set the chitted potatoes in place with the sprouts uppermost, and gently cover them with a ridge of soil to a depth of about 5–6 in.

EARTHING UP

When the foliage appears – called the haulm – take a hoe and gently draw the earth up around the plants until all but a few inches are covered. Repeat this 'earthing up' every few days or so during the growing season, until you finish up with a ridge about 1 ft high.

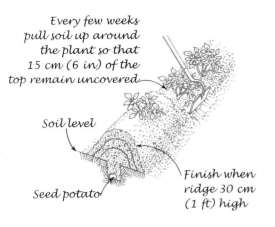

Every few weeks pull soil up around the plant so that 15 cm (6 in) of the top remain uncovered

Soil level

Seed potato

Finish when ridge 30 cm (1 ft) high

Caring for potatoes

NO-DIG CULTIVATION

Chit the seed potatoes as already described. In mid-spring, nestle the chitted potatoes directly in place on prepared ground, and cover them with a low ridge or mound of earth. Cover the ridge with black plastic sheeting and secure it by burying the edges in the ground. As soon as you see evidence of the potatoes pushing up under the plastic, take a sharp knife and very carefully cut a cross through the plastic so as to reveal the shoots.

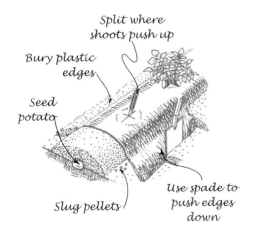

Split where shoots push up

Bury plastic edges

Seed potato

Slug pellets

Use spade to push edges down

Non-cultivation system

HARVESTING

Earlies In early to mid-spring or mid to late summer, depending on the variety, wait until the flowers are fully open, and then use a fork to gently lift and ease the potatoes from the ground.

Maincrop In early to mid-autumn, wait until the tops have completely died down and fork them up as already described. With the no-dig system, you can simply roll back the plastic sheet and take potatoes as and when needed.

STORING

Potatoes can be stored in a clamp – that is piled up in a dry corner of the garden and then covered with straw and buried – or they can be put in sack and stored away in the dark in a dry frost free shed.

TROUBLESHOOTING

Potato blight Brown, rust-like patches appear on the leaves; also sometimes white mould on the underside. There is not much you can really do about this disease, other than to burn the plants, grow on another plot the next time around, and use a disease-resistant variety.

Slugs Large, soggy, gray holes run through the potatoes, and there is plenty of slime, at its worst in late summer. Avoid growing potatoes in wet soil and to avoid digging in too much manure – there is nothing slugs love more than a low-lying, boggy area with lots of wet manure.

Cyst eelworms Brown, withered leaves and colonies of minute, ball-like cysts on the roots. Avoid growing potatoes on the same plot, only use resistant varieties, and encourage natural predators by following the no-dig system.

BRASSICAS

The botanical name for cabbage, *Brassica,* has come to be used by gardeners to describe a whole group of vegetables that includes Brussels sprouts, kale, broccoli, cauliflower and many others. Although the term 'brassicas' is also used in most textbooks to describe non-leafy items such as rutabagas and turnips, when most gardeners talk about brassicas they generally mean 'greens.' Traditionally, especially in the UK, a good part of Europe, the USA and Australia, greens are a most important staple. If you get it right, it is easily possible to have a steady, year-round supply of greens.

Brassicas in their various forms (such as broccoli and cabbage) are one of today's 'miracle' foods: they contain iron, calcium and vitamins C and E, and are low in fat and high in sodium. For this reason, they ought to figure highly in your self-sufficient plans.

When it comes to growing them, although there are slightly different methods, the overall procedures more or less follow the same pattern of sowing and thinning out, or sowing and transplanting, and then general care.

GROWING METHOD 1 – SOWING IN BEDS AND TRANSPLANTING

Prepare the bed as for any other vegetable – raking and firming until you have a fine tilth. Set out lines about 9 in. apart, make drills about ½ in. deep, sow a thin line of seeds, rake to cover, and water with a fine spray. When the seedlings are large enough to handle, thin out to leave the strongest seedlings about 2 in. apart. When the seedlings are good and strong, water the ground, and then lift and transplant them to their final growing positions.

GROWING METHOD 2 – SOWING AND THINNING

Prepare the bed by raking until you have a fine tilth. Set lines about 18 in apart and sow a pinch of seeds in ½ in deep drills about 9–12 in apart. When the seedlings are big enough to handle, thin them out to leave one strong plant at every station. When the plants are nicely hearting up, pick and eat every alternate

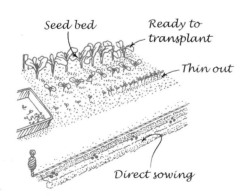

Seed bed — *Ready to transplant* — *Thin out* — *Direct sowing*

Sowing brassica seeds

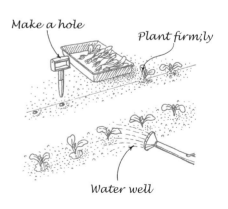

Make a hole — *Plant firmly* — *Water well*

Transplanting brassicas

plant as spring greens, leaving one good plant at every 14–24 in, depending on variety.

TRANSPLANTING

Prepare the ground by raking until you achieve a good tilth, and set lines 1–2 ft apart. Use the dibber (dibble) to make holes 1–2 ft apart – the spacing depends on the variety.

One at a time, set a seedling in the hole and flood it with water so as to 'puddle' the roots into the hole. Gently ease the seedling up to the right level, scoop earth to fill the hole, and then use your fingertips to firm the seedling into place. Water the row of seedlings in the early morning and late evening until they are well established.

WEEDING AND MULCHING

As soon as the weeds show, take the hoe of your choice and skim off the weeds just below soil level. If the weather is hot, dry-stir the surface of the soil so as to create a loose-soil mulch – a layer of broken soil that will reduce evaporation. Finally, cover the soil around the plant with a fine mulch of spent manure.

HARVESTING

Harvest year-round as and when needed.

TROUBLESHOOTING

Wind damage Weak-looking plants wobble on their stalks. The best prevention is to hoe the earth high up around the stems so that the relatively fragile stem is buttressed.

Insects Holes in leaves, colonies of eggs and/or grubs on the leaves, colored patches and so on. A good defence is to cover the young plants with fine-grade fleece, which will keep off butterflies, greenfly and other insects.

Birds Pecked and damaged leaves and flowers, and lots of soil disturbance around the plants. The simplest option is to use plastic netting to build a temporary barrier over the whole line. If this is too expensive, set up a ring of canes and run a crisscross pattern of black cotton over the whole bed.

Cabbage-white caterpillars In the first instance you will see white butterflies, then lots of minute eggs on the underside of the leaves, and finally caterpillars. The easiest solution is to wash the eggs off as soon as you see them, and/or cover the plants with fine netting or fleece.

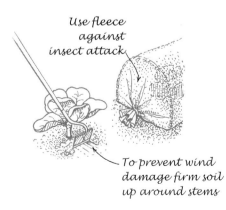

Keep free from weeds by regular hoeing

Mulch with well-rotted compost

Use fleece against insect attack

To prevent wind damage firm soil up around stems

Weeding and mulching brassicas

Summer insect attack and winter wind damage

BEETROOT

The beet family includes such vegetable delights as spinach, chard, beet leaf, spinach beet and one or two others, but the best-known member is beta or beetroot, grown for its plump, red, juicy roots. Beetroot has long been a popular vegetable, the evidence being that the word 'beta' is an ancient Celtic word meaning 'red.' People in the UK tend to eat it as a cold side dish with salad, and in the form of a pickle or relish, whereas diners in the USA and many parts of Europe favor it in the form of soup, and as a hot side dish to cooked meat.

From a self-sufficiency viewpoint, beetroot is a winner on many counts: it is a healthy food – no fat, few calories and lots of iron, potassium and vitamin C; it is easy to grow; it can be cropped in summer, autumn and winter; and it is easy to store in the form of dried or bottled roots, or as pickles and chutneys. When it comes to growing beetroot, the seeds are so large and manageable that they can simply be sown in place.

GROWING METHOD 1 – SOWING IN LINE

Prepare the ground as for any other vegetable, raking and firming until you have a fine tilth. Soak the seeds for about one hour. Set out lines about 12–15 in. apart, make drills about 1 in deep, sow a thin line of seeds, rake to cover, and water with a fine spray. When the seedlings are large enough to handle, thin out to leave the strongest seedlings about 4–9 in. apart. Do not forget to water generously at every stage.

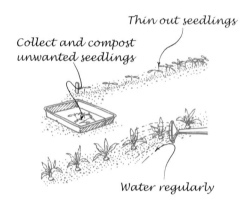

Thin out seedlings

Collect and compost unwanted seedlings

Water regularly

Thinning out and watering

GROWING METHOD 2 – SOWING IN STATION

Prepare the bed by raking until you have a fine tilth. Soak the seeds for one hour. Set lines about 12–15 in. apart and sow 2–3 seeds either in 1 in deep drills or in holes about 4–9 in. apart. When the seedlings are big enough to handle, thin them out to leave one strong plant at every station. The advantage of this method is that you save on seeds and minimize disturbance.

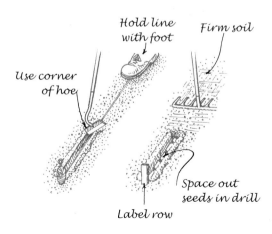

Hold line with foot *Firm soil*

Use corner of hoe

Space out seeds in drill

Label row

Planting beetroot

GROWING METHOD 3 – BROADCAST SOWING IN BEDS

Prepare the bed – best if it is raised and about 4 ft square – and rake until you have a fine tilth. Soak the seeds for one hour. Scatter the seeds over the ground and spend time carefully spacing them so that they are all about 6 in. apart. Water the bed with a fine spray and cover the whole bed with a sheet of plastic. Remove the plastic as soon as the seedlings show. The main advantages of this method are that the space is used to best effect, the seedlings are not disturbed, and the plants are so close that the leaf cover acts as a mulch to hold in water and restrict the growth of weeds.

WEEDING AND MULCHING

As soon as weeds show, take the hoe of your choice – I like using a very small, single-handed, swan-necked hoe – and skim off the weeds just below soil level. Make sure that you do not scuff, cut or in any way damage the shoulders of the beetroot. If the weather is hot and dry, stir the surface of the soil so as to create a loose-soil mulch – a layer of broken soil that will help reduce the evaporation of

Carefully remove the weeds

Remove dead or decaying leaves

Weeding and clearing dead leaves

residual water. Finally, cover the soil around the plant with a fine mulch of spent manure. Remove any wilting or yellowing leaves.

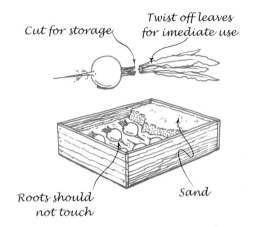

Cut for storage

Twist off leaves for imediate use

Roots should not touch

Sand

Storing beetroot

HARVESTING AND STORING

Lift and eat from early summer. To store beetroot, in mid to late autumn carefully lift the roots, twist off the leaves, and store the roots in boxes of sand or peat, in a single layer or in multiple layers, depending on the depth of the box.

TROUBLESHOOTING

Beet and mangold fly Maggots on the lower leaves or upper shoulders of the root. The easiest and most direct solution is to pick off and destroy affected leaves as soon as you spot a problem.

SEAKALE AND SPINACH

These two other members of the beet family are grown in much the same way as beetroot. The only real difference is that for seakale and spinach you concentrate your efforts on cultivating the leaves rather than the roots.

LEGUMES

The family of plants known as legumes includes runner beans, broad beans, butter beans, French beans and peas. Legumes are not only a primary food source that can be used as the basis for a balanced diet – they are low in fat, high in protein, iron and fiber, and very good for the heart and general muscle development – but they also enrich the soil with nitrogen. Legumes are beneficial in many ways: they are important as a food, relatively easy to grow, offer lots of variety and are easy to store.

GROWING RUNNER BEANS

Some growers sow seeds directly into open ground between late spring and early summer, others opt for an early start by raising seedlings under glass in mid- to late spring, and some do both. Traditionally, growers spend as much time building supports as actually tending the crop. Runner beans do best in a well-prepared, middle to light loam. Clay soils are just about acceptable if they are open and well drained, but overall they are least suitable.

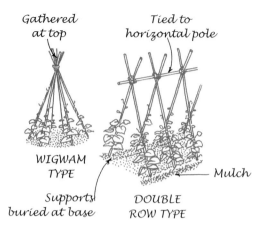

Gathered at top

Tied to horizontal pole

WIGWAM TYPE

Mulch

Supports buried at base

DOUBLE ROW TYPE

Plant supports

Prepare a trench, 2 ft wide and 9 in. deep, as early in the year as conditions allow. Fill it with rotted manure topped with soil. Build a support frame. To raise seedlings under glass in mid- to late spring, plant individual seeds in compost-filled pots. To sow seeds directly in the ground in late spring to early summer, run 2 in. deep drills along the trench and sow the seeds 5–6 in. apart. Plant the seedlings out at the end of spring. Set them at 5–6 in. intervals along the prepared trench.

PLANT CARE

Stir the soil with a hoe to create a loose-soil mulch to keep it free from weeds. In dry weather, spread a mulch of straw or old manure at either side of the row to help hold in moisture.

HARVESTING

Harvest from mid-summer to mid-autumn when the pods are young and slender, just at the point when the shape of the beans within the pod begins to show.

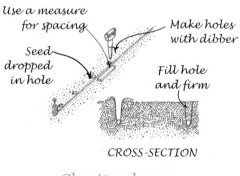

Use a measure for spacing

Make holes with dibber

Seed dropped in hole

Fill hole and firm

CROSS-SECTION

Planting large legume seeds

GROWING BROAD BEANS

Broad beans will grow in almost any kind of soil, but do best in a deep, rich, moisture-retentive soil that has been enriched with well-rotted farmyard manure. Experienced growers favor planting in late autumn and cropping in early summer. As a general guide, broad beans like a sunny, sheltered spot with a moist, well-drained soil, and dislike draughts and waterlogged, heavy soil. In late winter to mid-spring for maincrop beans, or late autumn for a winter-hardy crop, sow single seeds in 3 in. deep drills, or 3 in. deep dibbed holes, about 5–8 in. apart in rows 18 in. apart. Water generously.

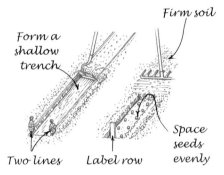

Firm soil

Form a shallow trench

Two lines   *Label row*   *Space seeds evenly*

Planting medium legume seeds

PLANT CARE

When the plants are about an inch high, draw soil up around the stems. Stir the ground with a hoe to provide a loose-soil mulch. As soon as the blooms are set, pinch out the top shoots to plump up the pods and deter blackfly. Support with sticks and string.

HARVESTING

Harvest between early summer and mid-autumn when the beans are firm. Tweak individual pods from the stem. When the crop is finished, cut the plants down to ground level to leave the nitrogen-rich roots in the soil.

GROWING PEAS

Early crops favor a warm, dry, sandy soil, while main crops prefer a heavier, richer, moisture-retentive loam. Peas need lots of moisture. If you see the ground cracking or drying out, soak it with water and follow by covering the ground. Sow seeds, in succession from early spring through to mid-summer, in 2 in. deep drills, at 5–6 in. intervals, in rows 15–48 in. apart, depending on the height of the variety. Cover the rows with something like wire mesh, twigs or cotton threads – anything to keep the birds away – and build a support frame with netting.

PLANT CARE

As soon as the row of seedings is up, stir the surface of the soil with a hoe to create a fine loose-soil mulch. Repeat this hoeing at least every week throughout the season, especially in dry weather. In long dry spells, drench the soil with water and put a heaped line of old manure mulch on each side of the row.

HARVESTING

Harvest from early summer to mid-autumn. Pick the pods every 2–3 days to encourage new pods to develop.

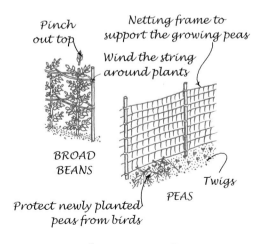

Pinch out top   *Netting frame to support the growing peas*

Wind the string around plants

BROAD BEANS

Protect newly planted peas from birds   *PEAS*   *Twigs*

Plant supports

ROOT VEGETABLES

In the context of this book, root vegetables or roots are simply vegetables where we eat the part that grows below ground. Roots like a moist, well-manured soil, do not do well in stony soil, are generally sown from seed in their final growing positions, and dislike being transplanted. Carrots look good, smell good, taste good, do you good and are relatively easy to grow; most kids like them, there are lots of varieties, and, best of all, they can be eaten from early summer through to early winter. Parsnips are hardy, can be grown in just about any well-prepared, fertile soil, and can be left in the ground over winter. Salsify and scorzonera do not look pretty on the plate, but they crop heavily and are very tasty.

GROWING CARROTS

Carrots do best on a friable, deeply dug, well-drained, sandy, fertile loam in a sunny position. If your soil is overly stony or contains great lumps of fresh manure, the chances are that the growing root will divide and become stunted or forked. Sow seeds from early spring to early summer in ¾ in. deep drills, 6–9 in. apart. Sow thinly and water with a fine spray. Thin the seedlings so they are about 2 in. apart. Use cloches to protect the plants at the beginning and end of the season.

PLANT CARE

Hoe frequently to provide a loose-soil mulch and to deter weeds. Water little and often.

HARVESTING AND STORING

You can harvest from late spring through to early winter, if you have made successive sowings of a range of varieties, and if you protect the crops so that they can stay in the ground. Store maincrop carrots in boxes of sand in a frost-free shed.

GROWING PARSNIPS

Parsnips do best on just about any well-prepared, friable, fine-textured, well-drained, fertile soil that has been manured for a preceding crop. Avoid using fresh manure, because if the growing roots hit fresh manure they will either fork or become cankered. Dig the soil to a good depth, so that the roots can

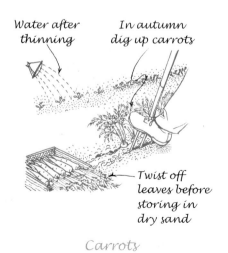

Water after thinning

In autumn dig up carrots

Twist off leaves before storing in dry sand

Carrots

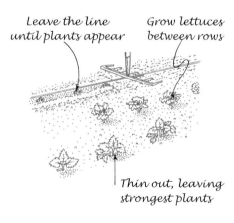

Leave the line until plants appear

Grow lettuces between rows

Thin out, leaving strongest plants

Parsnips

go straight down without obstruction. If you have to grow parsnips on poor, stony soil, choose a short, stubby variety. Sow in late winter to mid-spring in ¾ in. deep drills, 8–10 in. apart. Sow a pinch of 3–4 seeds at 5–6 in. intervals. Firm the soil and water generously. When the seedlings are big enough to handle, thin to leave the strongest plant at each station.

PLANT CARE

Stir the surface of the soil with a hoe to create a loose-soil mulch. Be very careful that you do not scuff or graze the emerging root; such damage often results in top-rot or canker. Water little and often to avoid wet-drought-wet conditions that result in root splitting.

HARVESTING

Harvest from mid-autumn to early spring. They are frost-hardy, so you can leave them in the ground until needed. Ease them up with a fork, being careful not to break the root tips.

GROWING SALSIFY AND SCORZONERA

Salsify and scorzonera do best in a deep, light-textured, moist, friable, stone-free, fertile soil. The soil must be well prepared with lots of well-rotted manure. Avoid digging in fresh manure, as it will cause the roots to divide and possibly rot, and also result in the flavor being a bit rank and muddy. Sow seeds in mid- to late spring in ½ in. deep drills, 10–12 in. apart. Sow a pinch of 2–3 seeds at 5–6 in. intervals. Firm the soil and use a fine spray to water generously. When the seedlings are large enough to handle, thin to leave the strongest plant at each station. Water before and after thinning, and firm the earth up around the remaining plants.

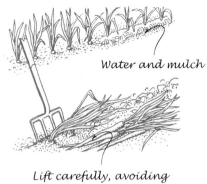

Water and mulch

Lift carefully, avoiding damaging the root

Salsify care

PLANT CARE

Stir the surface of the soil with a hoe to create a loose-soil mulch. Take care not to scuff the emerging shoulder; such damage often results in top-rot or canker. Water little and often to avoid root splitting.

HARVESTING

Harvest from mid-autumn to mid-spring. Lift and eat as needed. A good option with a large crop is to chop the old tops off in autumn, hoe soil over to a depth of 5–6 in., and then eat the resulting blanched shoots ('chards') in the following spring.

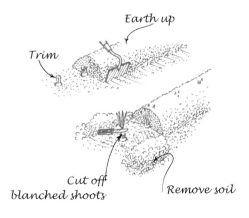

Earth up

Trim

Cut off blanched shoots

Remove soil

Producing chards

ONIONS

Members of the family of plants that includes onions, garlic, leeks and shallots have been described as some of today's 'miracle' foods. They contain the antioxidant quercetin and a whole batch of other micronutrients that are proven to reduce blood clotting, they are good for reducing high cholesterol levels, and they are low in calories and high in vitamins B and C. They are also relatively easy to grow, they can be stored in a whole variety of ways and, most importantly, they are tasty.

Leeks are tended in much the same way as celery, endives and chicory, in that the plant needs to be blanched. Leeks are a good option for winter use, when other vegetables are scarce. Varieties range from small and mild through to large and strong-tasting. Choose a variety to suit both your taste and your soil.

GROWING ONIONS

Onions do best on a deeply worked, well-manured, moist, well-drained, friable, fertile, sandy soil in a sunny position. Ideally the manure needs to be well broken down and rotted. To this end, it needs to be dug in during the autumn so that the bed is ready for spring sowing. Almost any kind of manure will do as long as it is well rotted. Some reckon that for good large onions you cannot do better than spread a topdressing of dried chicken manure and sand. If your soil is sandy it needs a dressing of ground clay, and if it is heavy and sticky it needs a dressing of sand or grit.

GROWING ONIONS FROM SEED

Sow seeds thinly in early spring or late summer in ½ in. deep drills, 10 in. apart. Thin the seedlings to 1–2 in. apart, depending on the variety.

Thin when seedlings are upright

Seed

Push set into soil leaving the nose exposed

Keep free of weeds

Planting onions

GROWING ONIONS AND SHALLOTS FROM SETS

Plant sets in late winter to mid-spring 2–6 in. apart, in rows 10 in. apart.

PLANT CARE

Stir the surface of the soil with a hoe to create a loose-soil mulch. Later use your fingers to draw the soil slightly away so that the swelling bulb sits high on the surface. Do not water, but rather keep stirring the ground

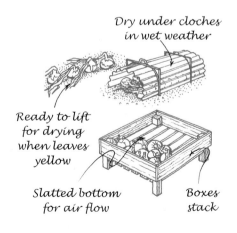

Dry under cloches in wet weather

Ready to lift for drying when leaves yellow

Slatted bottom for air flow

Boxes stack

Ripening onions

with the hoe to prevent the moisture wicking out from the underlying moist soil.

HARVESTING AND STORING

Harvest from early summer to mid-autumn. Lift salad onions, and large onions, as needed. To ripen and store, first bend the tops over. When the tops are yellow, lift the onions and lay them in the sun to dry. Finally, put them in boxes, or string them up, and store in a frost-free shed.

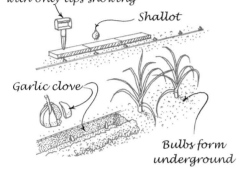

Make a hole and plant with only tips showing

Shallot

Garlic clove

Bulbs form underground

Planting shallots and garlic

GROWING GARLIC

In late spring, plant single cloves in 1 in. deep drills, 8 in. apart, with 1 ft between rows. Water frequently.

PLANT CARE

Stir the surface of the soil with a hoe to create a loose-soil mulch and to remove weeds. Once the plants are well established, ring them around with a thin mulch of old spent manure.

HARVESTING

Harvest from summer onwards, the precise time depending on your chosen variety and growing methods. Use a fork to help ease the clumps from the ground, and lift them as needed.

GROWING LEEKS

Leeks can be grown on just about any soil, but prefer it to be deeply worked, well manured, moist and well drained. Dig holes or trenches and set the leeks in a mix of old well-rotted manure and compost. The soil kept ready for earthing up must be repeatedly worked so that it is fine, slightly moist and friable.

For main crops (to harvest from early autumn to mid-spring), sow thinly from early to mid-spring in ½ in. deep drills, 10–12 in. apart. When the seedlings are up, transplant them into 6 in. deep holes at 6 in. intervals, with 1 ft between rows. Place one seedling in each hole, and top up with water.

PLANT CARE

Hoe to create a loose-soil mulch. Once the transplanted seedlings are established, ring them with a mulch of chopped straw or old spent manure, and then spend time every week with the hoe drawing the earth up. Repeat throughout the season until all but the tops are covered. Water frequently.

HARVESTING

Harvest from early autumn to late spring, depending on variety and growing methods. Use a fork to ease the roots from the ground.

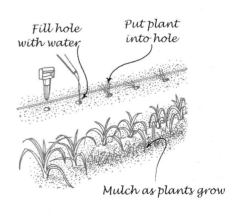

Fill hole with water

Put plant into hole

Mulch as plants grow

Planting leeks

CUCURBITS

Cucurbits is the botanical name for the annual climbing or trailing plants of the gourd family, including marrows, courgettes (zucchini), cucumbers, gherkins, pumpkins, melons and summer squashes. Although the flesh is mostly water, it is still a good source of beta carotene, vitamin C and folate, and low in calories. The seeds are high in vitamin B, unsaturated fats and protein. Cucurbits are dramatic and tasty.

GROWING MARROWS AND COURGETTES (ZUCCHINI)

While they can be grown almost anywhere, they do best in deeply worked, well-manured, moist, well-drained soil. A good method is to dig holes or trenches and build a heap of well-rotted manure and compost. Fill a hole with manure, cover it with a flat mound of manure topped with about an inch of earth, and then insert the plants. The mound catches the sun and ensures that the soil is well drained. Once the plants are growing well, cover the mound with a mulch to hold in the moisture.

Sow seeds in peat pots in mid to late spring – two seeds per pot – and place them under glass or plastic. When the seedlings have two leaves, carefully thin to one good plant in each pot. In late spring to early summer, when the stems are hairy, dig holes 1 ft deep and wide, in rows 4–5 ft apart, and fill them with well-rotted manure. Build a soil-manure mound over each hole. Set the peat pots into the top and water.

PLANT CARE

Hoe to create a loose-soil mulch. Spread a mulch of spent manure around the plant. Pinch the growing tip out at about 2 ft. Keep watering and mulching through the season with grass clippings and more spent manure to plump up the crop.

HARVESTING AND STORING

Harvest from mid-summer to mid-autumn. Cut courgettes every few days to keep them producing, and marrows as needed. At the end of the season, wait until marrows are hard-skinned and a good size, and store them in nets in a frost-free shed.

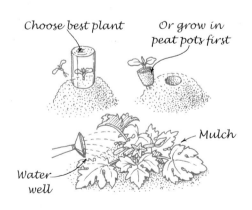

Planting marrows and courgettes

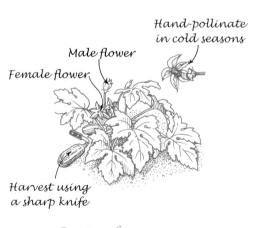

Caring for marrows and courgettes

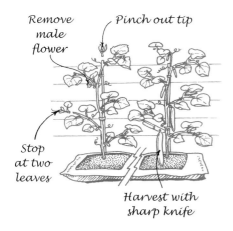

Cucumbers indoors

Remove male flower
Pinch out tip
Stop at two leaves
Harvest with sharp knife

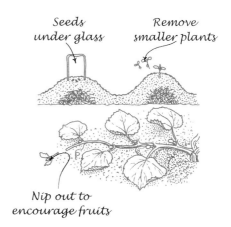

Cucumbers outdoors

Seeds under glass
Remove smaller plants
Nip out to encourage fruits

GROWING CUCUMBERS

If you enjoy growing marrows and courgettes (zucchini), you will probably also like growing cucumbers. You have the choice of growing them under cloches, in a cold frame, or outside with some sort of protective screen placed on the windward side. Make sure that you choose a variety to suit your particular needs – for example, if you are growing them out of doors, make sure that you choose ridge or gherkin types. Growing-bags are a very good option as you can sow the seeds directly in the bags.

The soil needs to be light to medium, soft in texture, deeply dug, well manured and moist. For outdoor varieties, dig a pit or trench depending on how many plants you are planning to grow, and top the pit or trench up with a mix of compost and/or well-rotted manure. The aim is to have a moist but well-drained soil that is rich in humus.

Choose a spot in full sun – well away from draughts – and put a plastic screen or shelter on the windward side. In earlier times, gardeners used to plan out the plot so that the area was protected from cold winds by rows of peas and beans.

GROWING METHOD 1 – INDOORS

Sow the seeds in late spring in trays (flats) on moistened tissue, cover with glass and keep warm. As soon as the seedlings are large enough to handle, prick them out into 3 in. peat pots and later into 9 in. pots or growing-bags. When the plants are established and growing in height, support them with canes and ties. Pinch out the growing tip when it reaches the top of your greenhouse or cloche.

GROWING METHOD 2 – OUTDOORS

In late spring dig trenches 1 ft deep and wide, and fill with manure and soil. Sow three seeds ¾ in. deep, 2–3 ft apart, and cover with plastic bottles. When the seedlings are up, thin to the best plant at each station and cover with glass or plastic. Let the plant trail along the ground. Pinch out the tips of the sideshoots when they reach the frame or boundary, and water daily.

HARVESTING

Harvest indoor varieties from mid-summer to mid-autumn, and outdoor varieties from early summer to early autumn. Support the weight of the fruit and cut off with a sharp knife.

FRUITING VEGETABLES

'Fruiting vegetables' are fruits grown in the vegetable plot that are treated as vegetables. So, while tomatoes, rhubarb and strawberries are all fruits, only tomatoes are a fruiting vegetable. Outdoor tomatoes tend to be sturdier, healthier, and tastier than those grown indoors. Greenhouse tomatoes can be planted earlier and cropped later than outdoor varieties, but they are more prone to diseases. Capsicums (sweet peppers) and eggplants like similar growing conditions. You can sow them directly in growing-bags, or in seed trays (flats) and transfer to peat pots and on to large pots – outdoors or under glass. New varieties of sweetcorn can, with care and protection, be grown in a cool climate.

GROWING OUTDOOR TOMATOES

Outdoor tomatoes prefer a deeply worked, richly manured, compact, well-drained soil in a sunny, sheltered position, with protection on the windward side. In early to mid-spring, sow in trays (flats) on moistened potting compost, and protect with a sheet of glass. In mid to late spring, when seedlings are large enough to handle, prick them out into 3 in. peat pots.

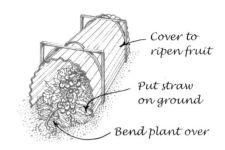

Cover to ripen fruit

Put straw on ground

Bend plant over

Ripening outdoor tomatoes

In late spring to early summer, set the peat pots in a sheltered position and water.

PLANT CARE

Support the plant with stakes and ties. Pinch out the sideshoots. Remove the growing tips when there are 5–6 trusses. When the fruit starts to ripen, cover the ground with a mulch of well-rotted manure topped with a bed of straw, remove the supports, and gently lay the plants down. Cover with the cloche and water the roots. When the fruits start to ripen, strip off all the leaves up to the first leaf above the bunch, and remove damaged fruits. Check that the straw is crisp and dry.

HARVESTING

Harvest in mid-summer to mid-autumn when the tomatoes are firm and nicely colored.

GROWING GREENHOUSE TOMATOES

The easiest option is to plant them in growing-bags. In late winter to early spring, sow seeds in trays (flats) of moistened potting compost, and protect with a sheet of glass. In mid to late spring, when the seedlings are large enough to handle, prick them out into the growing-bags and water generously.

Remove sideshoots

Nip out growing tip two leaves above last flower truss

Flower truss

Tomatoes

PLANT CARE

Support the plant with stake and ties. Pinch out sideshoots that grow from the crutch between leaf and stem. Remove the growing point just short of the top of the greenhouse. Water little and often, with a dedicated liquid feed in the water. Remove yellow foliage.

HARVESTING

Harvest from early summer to mid-autumn, depending on variety. When the first fruits begin to ripen, strip off all the leaves from the ground to the first leaf above the bunch. Pluck the tomatoes when firm and nicely colored.

GROWING CAPSICUMS (SWEET PEPPERS) AND EGGPLANTS

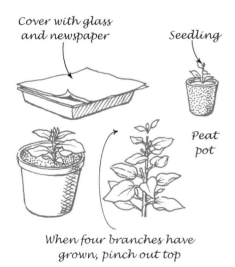

Cover with glass and newspaper

Seedling

Peat pot

When four branches have grown, pinch out top

Sweet pepper

Peppers and eggplants prefer a well-drained, well-manured soil in a sunny, sheltered position. In late winter to early spring, sow the seeds in trays (flats) on moistened potting compost, and protect with a sheet of glass. Keep warm. In mid to late spring, prick the seedlings out into 3 in. peat pots. In early summer, when the plants are strong enough to handle, either set the peat pots in 9 in. pots or growing-bags and put them in the greenhouse or polytunnel, or plant them outside in a prepared bed.

PLANT CARE

Water the seedlings before and after planting. Hoe to create a loose-soil mulch and to remove weeds. Support the plants with canes and ties. If the weather is dry, spread a mulch of spent manure or straw over the ground to hold in the moisture. Water frequently. When the plant is about 1 ft high, remove the growing tip. Remove all but the best six fruits and subsequent flowers.

HARVESTING

Harvest from mid-summer to mid-autumn, depending on whether you want green or red peppers, or small or large eggplants. Use a sharp knife to cut the fruits as needed.

GROWING SWEETCORN

Sweetcorn prefers a well-prepared, deeply worked, well-drained, light to sandy soil in a sheltered, sunny position. Dig in plenty of manure and compost in autumn. In a cool climate, choose a variety like 'Northern Xtra Sweet.' In mid to late spring, sow one chitted seed per peat pot under glass or plastic and water. In late spring to early summer, dig holes 12–18 in. apart in a grid pattern, set the pots in place and water generously.

PLANT CARE

Hoe to create a loose-soil mulch. Drag the soil up to support the stems. In dry weather, when the flowers are over, water and cover the ground with a thick mulch.

HARVESTING

Harvest from mid-summer to mid-autumn when the tail-like silks are black-brown. Pick with a swift twist-and-down action.

SALAD LEAVES

In this book, 'salad leaves' are just leaves that you can pick and eat raw, whether tossed in oil or stuffed into a cheese sandwich. There are many easy-to-grow options: chicory, endives, lettuces, chives, land cress, lamb's lettuce, some of the herbs – they are all good.

If you enjoy a crisp, tart flavor and you have a greenhouse or cold frame and the use of a shed, and if you do not mind waiting a good part of the year for a crop, chicory is a good option. The bitter, 'smoky burnt' taste is perfect with something like a crumbly, white, Welsh cheese. Endives can be harvested from late summer to the following spring – perfect for winter eating. If you get it right, you can follow on spring and summer lettuces with endives. They need to be blanched, which creates a sweet, but slightly tart taste. Some of the large, bushy-head varieties are self-blanching. Using hotbeds, fleece, cloches and cold frames, you can be eating lettuces right through the year. If you grow lettuces intercropped with radishes, you can grow a pretty good salad all in one go.

GROWING CHICORY

Chicory prefers a light to medium, deeply dug, fertile, moist, friable soil in a sheltered, sunny position. Sow seeds in late spring to mid-summer in ½ in. deep drills, 10–12 in. apart. Sow thinly and water with a fine spray. Thin the seedlings to about 8 in. apart.

PLANT CARE

Water daily. Stir the surface of the soil with a hoe to create a loose-soil mulch. When the leaves have died down, lift and trim the roots, and bed them in dry sand.

Trim shoots and roots

Exclude all light with tape

Chicory

HARVESTING

Harvest and eat self-blanching varieties from mid-autumn to early winter. Blanch traditional types from late autumn to mid-spring. To do this, in late autumn, lift plants, cut off the leaves fairly close to the roots, and store the roots. When ready, place the roots, four at a time, vertically in a box of moist sand and cover with black plastic. The leaves will form white buds or 'chicons,' ready to eat in 3–4 weeks.

GROWING ENDIVES

Endives prefer a deeply dug, fertile soil that contains plenty of well-rotted manure from a previous crop. Varieties have slightly different sun-and-shade needs, so choose a sunny, open position and use screens to provide shade. Sow seeds in early spring to early autumn in ½ in.

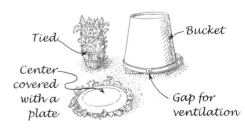

Tied

Bucket

Center covered with a plate

Gap for ventilation

Three methods of blanching

deep drills, 10–12 in. apart. Sow thinly and water generously. Thin the seedlings to 1 ft apart.

PLANT CARE

Hoe to create a loose-soil mulch, and spray daily. Blanch the plants between mid-summer and mid-autumn, when they are fully grown. To do this, on a day when the air is dry and warm, gently bind the plants up with raffia, and cover with a clay flowerpot. Check regularly to make sure they are dry and free from slugs.

HARVESTING

Harvest after the second week; lift the pot and cut the plant off close to the ground.

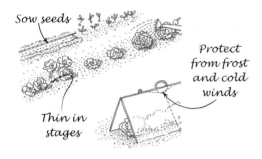

Sow seeds

Thin in stages

Protect from frost and cold winds

Lettuces

GROWING LETTUCES

Lettuces prefer a light, friable, well-drained, moisture-retentive, sandy loam. For a spring crop, sow seeds in late summer to mid-autumn in ½ in. deep drills, 10–12 in. apart. Sow thinly and use a fine spray to water. Protect with glass or plastic. Thin seedlings to 1 ft apart. For a summer crop, sow seeds in early spring to mid-summer in drills as above – but 8–9 in. apart for dwarf varieties. Thin seedlings to 9–10 in. apart so that the lettuces are in staggered rows.

PLANT CARE

Hoe to create a loose-soil mulch. Hoe and water right throughout the season.

HARVESTING

Harvest all year round, the precise time depending on your varieties. Pick and come again, or cut close to the ground as needed.

CHIVES
Remove flowers and divide every few years

PARSLEY
Protect for winter harvesting

Chives and parsley

OTHER OPTIONS

Chives – A hardy, low-growing perennial, with green tubular stems topped with round, rose-pink flowerheads. The chopped-up stems have a beautifully subtle, onion-like flavor – very good in a tossed salad.

Fennel A hardy herbaceous perennial with tall stems, feathery green leaves, and golden-yellow flowerheads that grows to a height of about 5–6 ft. The leaves are particularly good tossed in olive oil and eaten with rye bread.

Lamb's lettuce A good substitute for lettuce, very easy to grow.

Land cress Not just a stand-in for water-cress, but a tasty item in its own right; it has a peppery taste, and makes a good sandwich.

Parsley A hardy biennial that tends to be grown as an annual, with curly, tightly packed green leaves. Good in a mixed-leaf salad.

Rocket A hardy, spicy-tasting, green-leaf plant that is good as a cut-and-come-again salad.

Watercress A good choice if you have a source of free-flowing fresh water, easy to grow, nutritious and tasty.

STEM VEGETABLES

We use the term 'stem vegetables' to describe common vegetables like celery, celeriac, asparagus and kohl rabi. Celeriac and kohl rabi do look to be more root than stem, but the fat, root-like part is technically a stem. Celery is a valuable, high-fiber, low-calorie food that tastes good raw and in soups. Celeriac makes a delicious winter soup and is good in a winter salad. The good thing about kohl rabi is that from plot to plate only takes 6–8 weeks. It is delicious when steamed and mashed with butter, or eaten raw in a salad. Asparagus might take 3–4 years to establish, but it then crops for 10–20 years. The swiftest option is to grow it from one-year-old crowns.

GROWING CELERY

Celery prefers a deep, rich, moist, well-drained soil, in an open, sunny position. Soil used for earthing up must be fine and friable. Sow under glass in early to mid-spring. Prepare a trench 1 ft deep, 22 in. wide, and put manure topped with earth in the bottom 9 in. Prick the seedlings out into peat pots or trays (flats), 2 in. apart. Later set the plants in the trench, 9 in. apart.

PLANT CARE

Hoe around the plants to create a loose-soil mulch, and remove any suckers and dead leaves. When the plants are about 10–12 in. high, bind them with raffia, and earth up to cover all but the topmost foliage. In frosty weather, protect the foliage with chopped straw.

HARVESTING

Harvest self-blanching varieties from late summer to late autumn, and traditional varieties from late autumn to spring.

GROWING CELERIAC

Celeriac prefers a rich, fertile soil, in a sunny, sheltered position, where the manure has been dug in during the preceding winter. Sow seeds in early to mid-spring in prepared seed trays (flats) under glass. Pot the seedlings on as soon as they are large enough to handle, into peat pots or trays, 2 in. apart. Plant out the seedlings, in late spring to early summer, 1 ft apart, with 10–15 in. between rows, in shallow holes, so that they sit on rather than in the ground.

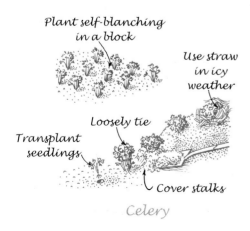

Plant self-blanching in a block

Use straw in icy weather

Loosely tie

Transplant seedlings

Cover stalks

Celery

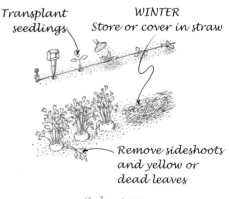

Transplant seedlings

WINTER Store or cover in straw

Remove sideshoots and yellow or dead leaves

Celeriac

PLANT CARE

Stir the surface of the soil with a hoe to create a loose-soil mulch, and to keep the ground free from weeds and bugs. Remove old leaves, sideshoots and roots as soon as the need arises. Never allow the ground to dry out.

HARVESTING AND STORING

If your site is well drained and you protect the plants with straw or fleece, you can leave the crop in the ground and harvest from late autumn to early spring; if the soil is wet, however, you must lift them and store in a frost-free shed.

GROWING ASPARAGUS

Asparagus prefers a well-drained, deep, rich soil that is inclined to sandy, in a sheltered, sunny spot. The ground should be deeply dug

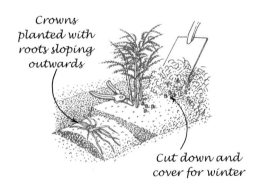

Crowns planted with roots sloping outwards

Cut down and cover for winter

Asparagus second year

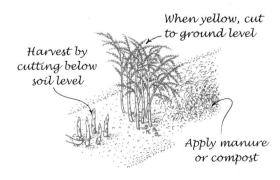

When yellow, cut to ground level

Harvest by cutting below soil level

Apply manure or compost

Asparagus third year

and well prepared with lots of well-rotted farmyard manure. Dig a trench 10 in. deep and 15 in. wide. Cover the base with 3 in. of mounded compost or well-rotted manure. In early to mid-spring, set the crowns 18 in. apart, with the roots spread over the mound, and cover with 2–3 in. of soil. Water generously.

PLANT CARE

In the first and second autumns after planting, cut the foliage, hoe to provide a loose-soil mulch, and cover with well-rotted manure topped with soil. In the third autumn, and in all future years, repeat the procedure, but let the spears grow. Rake off loose soil in spring.

HARVESTING

In the third season harvest from mid-spring to mid-summer. When the spears are 3–4 in. above ground, take a long-bladed bread knife and slide the blade at a flat angle into the soil to sever the spear at a point 3–4 in. below ground level.

TROUBLESHOOTING

Asparagus beetle Grubs and beetles eat both stems and leaves. Pick off larvae and beetles and repeatedly wash plants with a diluted mild insecticidal soap solution. Remove yellowing leaves and damaged stems.

Rust Brown-red splotches show on asparagus leaves in early summer. Remove and burn suspect leaves and shoots.

Frost damage Protect by covering the plants with loose sacking, plastic netting, old newspapers or fleece – be careful not to crush or bend tender foliage.

Slugs The best option in a small garden is to remove by hand and destroy them. Set traps by leaving young lettuce leaves under old tiles. Some growers advocate growing a barrier of French marigolds around susceptible plants.

A–Z OF VEGETABLES AND SALAD CROPS

ARTICHOKES, GLOBE

FACTS/TIPS
- Plants grow from root suckers.
- They grow to a height of about 4 ft.
- Remove the first heads and crop the second season after planting.
- Plants reach their peak in the third or fourth year.

CALENDAR
Sow – early to mid-spring.
Plant – mid to late spring.
Harvest – early summer to mid-autumn.

SOIL AND SITE
- Prefers a deep rich soil in full sunlight.
- Needs plenty of moisture throughout the summer months, and a well-drained soil in the winter.
- Traditionally grown in beds.
- Plant one or more plants a year for four years, so as to have a succession of well-cropping plants.
- Dress with well-rotted farmyard manure and water generously.

SOWING AND PLANTING
- Propagate from tubers or root suckers.
- Plant the suckers singly about 4 in. deep, with 30 in. between plants.
- Tread in and water generously.
- Water daily until the plant looks well established.

PLANT CARE
In summer, stir the ground with the hoe to keep it loose-mulched and clean. In winter, cut the foliage down to the ground and cover the crowns with straw. In the second season, remove the straw and apply fertilizer/manure in spring.

HARVESTING
Harvest from early summer onwards, when the heads are mature but still closed. Cut them off with 2–3 in. of stem.

VARIETIES
Green Globe: Dark green heads with attractive foliage.
Purple Globe – Romanesco: Purple to red flower buds. Does well in long cold winters.
Violetta di Chioggia: Variety known as much for its beautiful purple-headed thistle-like good looks as for its taste.

ASPARAGUS

FACTS/TIPS
- Grow from one-year-old crowns and crop in the second/third year.
- A healthy plant will crop for about 10–20 years.

CALENDAR
Sow – early to late spring.
Plant – mid to late spring.
Harvest – in year 2–3, mid-spring to early summer.

SOIL AND SITE
- Prefers a well-drained, rich, well-dug soil that is inclined to be sandy.
- Does best when the soil is enriched with two buckets of manure to every square yard.
- Likes a sheltered but sunny spot.
- Hoe the soil to open it up and keep it free from weeds.

SOWING AND PLANTING
- Dig a trench 10 in deep and 15 in. wide. Cover the base with 3 in. of compost or well-rotted manure.
- Set the crowns 18 in. apart, with their roots spread over the compost, and cover with 2–3 in. of soil. Water generously.

PLANT CARE
In the first and second autumns after planting, cut the foliage, stir the surface with the hoe to provide a loose-soil mulch, and cover with well-rotted manure topped with soil. In the third autumn after planting, and in all future years, repeat the same procedure, but let the spears grow.

HARVESTING
In the third season, when the spears are 3–4 in. above ground, slide a long blade at a flat angle into the soil to sever the spear at a point 3–4 in. below ground.

VARIETIES
Connover's Colossal: Thick, fat, mid to light green stalks. Long-established variety that has proven its worth over the years.
F1 Jersey Knight: All-male, disease-resistant hybrid that produces good fat spears.
Mary Washington: Rust-resistant variety that produces an abundance of long, straight, thick and heavy, dark green to purple spears.

EGGPLANTS

FACTS/TIPS
- Best grown from seed, in much the same way as tomatoes.
- Grow indoors or outside under cover.

CALENDAR
Sow – late winter to early spring.
Plant – mid to late spring.
Harvest – mid-summer to mid-autumn.

SOIL AND SITE
- Does best in a deep, well-drained, fertile soil in a sunny, sheltered position.
- Dislikes frost, waterlogged soil and sharp winds.

- Grow in 9 in. flower pots filled with potting compost, or in growing-bags.
- Provide each plant with its own little individual plastic-sheet shelter.

SOWING AND PLANTING
- Sow the seeds in a tray on a bed of moistened potting compost, protect with a sheet of glass, cover with newspaper and keep warm.
- When the seedlings are large enough, prick them out into 3 in. peat pots.
- In mid- to late spring, set the peat pots in 9 in. pots, water generously, and place under cover.

PLANT CARE
When the plant is about 1 ft high, remove the growing tip to encourage branching. Support with a cane and ties. Pinch out to leave the best six fruits, and remove subsequent flowers. Spray with water to discourage aphids. Apply a liquid feed, and strip off the older foliage.

HARVESTING
In mid-summer to mid-autumn, when the fruits are 6–9 in. long, slice them off with a sharp knife.

VARIETIES
Black Enorma: Variety grown primarily for the size of its fruit.
Long Purple: Good traditional, tried-and-trusted, medium-early variety. Produces a violet colored fruit about 5–6 in. long.
Snowy: Early-maturing variety. Good choice if you live in an area where the summers are short and wet. Produces cylindrical white fruits about 7–8 in. long.

BEANS, BROAD

FACTS/TIPS
- Plants sown in late autumn can be cropped in early summer.
- Overwintered plants are more resistant to blackfly.

CALENDAR
Sow – late winter to late spring.
Harvest – early summer to mid-autumn.

SOIL AND SITE
- Prefers a deep, rich, moist soil that has been enriched with well-rotted farmyard manure.
- Likes a sunny, sheltered spot with a moist, well-drained soil.

SOWING AND PLANTING
- Late winter to mid-spring – for maincrop in sandy soil, prepare 3 in. deep drills; for maincrop in heavy soil, prepare 3 in. deep dibbed holes.
- Sow single seeds 5–8 in. apart in rows 18 in. apart.
- For winter crop, sow as above.
- Water generously; firm soil gently.

PLANT CARE
When the plants are about an inch high, draw the soil around the stems. Stir the soil to provide a loose soil mulch and to remove weeds. As soon as the blooms are set, pinch out the top shoots to hold back the blackfly. Water frequently and support with sticks and string.

HARVESTING
Depending on the variety and the planting date, harvest between early summer and mid-autumn. Tweak the firm pods from the stem. When the crop is finished, cut the plant to the ground to leave the root in the soil.

VARIETIES
Bunyard's Exhibition: English heirloom variety that produces a heavy crop with 9–10 long bean pods.
Dreadnought: Swift-growing variety tat produces good-sized pods with large white beans.
Witkiem Manita: Superior early variety to sow in autumn. Produces 5–7 good 7–8 in. long bean pods.

BEANS, FRENCH

FACTS/TIPS
- French beans can be cropped earlier than runner beans.
- Mature beans can be dried for winter use.
- You can lengthen the season by protecting the plants with fleece and/or plastic sheeting.

- They are a good option if you are short of space.

CALENDAR
Sow – mid to late spring, and early to mid-summer.
Harvest – early summer to mid-autumn.

SOIL AND SITE
- French beans do best on a light soil, away from winds and draughts.
- It is best if the ground has been deeply worked and heavily manured for a previous crop.
- Give a light dressing of lime before the seeds are sown.
- As soon as the plants are up, spread a mulch of old manure spread alongside the rows.

SOWING AND PLANTING
- Mid to late spring – for early crop prepare 2 in. deep drills 18 in. apart and sow the seeds 2–3 in. apart.
- Late spring to early summer: for maincrop, sow as above.
- Early/mid-summer: sow as above.
- Water generously.
- Early and late crops will need to be protected with cloches or polytunnels.

PLANT CARE
When the plants are a few centimeters high, draw the soil around the stems to protect against frost and draughts. Hoe and mulch with old spent manure and water generously at the roots. Support climbing varieties.

HARVESTING
Depending on the variety and the planting date, start picking from mid-summer onwards. When the beans are firm, twist them off cleanly.

VARIETIES
Barlotta Lingua di Fuoco: Swift-growing climbing variety that produces lots of green-pink pods.
Canadian Wonder: Traditional variety that produces a heavy crop of long, flat, bright green pods.
Sungold: Unusual-looking variety with slender, bright yellow beans.

BEANS, RUNNER

FACTS/TIPS
- Runner beans are one of the most popular vegetables.
- You can sow the seeds directly into the prepared ground and/or raise them under glass.

CALENDAR
Sow – mid-spring to early summer.
Harvest – mid-summer to mid-autumn.

SOIL AND SITE
- Runner beans prefer a middle to light loam.
- Clay soils must be open and well drained.
- Dig in lots of well-rotted farmyard manure.
- Some growers dig a 2 ft wide, 9 in. deep trench in autumn and use it as a compost pit.

SOWING AND PLANTING
- Prepare a trench 2 ft wide and 9 in. deep and fill with rotted manure topped with soil.
- Build the support frame.
- Raise seedlings under glass in compost-filled pots in mid- to late spring.
- Sow beans directly in the ground in late spring to early summer, 2 in. deep and 5–6 in. apart.
- Plant the seedlings out at the end of spring at 6 in. intervals.

PLANT CARE
Stir the soil with a hoe. In dry weather, spread a mulch of straw or old manure at either side of the row to help hold in moisture. Spray-water the plants in the evening. Pinch out the tips as soon as they reach the top of the frame.

HARVESTING
Harvest from mid-summer to mid-autumn. Pick when the pods are young and slender, when the shape of the beans begins to show. The more you pick, the more you get. Pick as often as possible. If you allow pods to grow to maturity, the plant will come to a halt.

VARIETIES
Prizewinner: Popular prizewinning variety that produces long, straight, dark to mid-green pods.
Red Rum: Heavy-cropping, disease-resistant variety that produces narrow, straight, 6–8 in. long pods.
Scarlet Emperor: Old favorite that produces a long, rough-textured bean – the perfect choice.

BEET LEAF

FACTS/TIPS
- Beet leaf is a good option if you cannot grow spinach.
- Pick when young and tender, and eat both the leaves and the stalks.
- Cook and serve the stalks in their own right, like asparagus or stir-fried celery.
- The colored-stemmed varieties look good in the flower borders.
- You can crop leaf beet right through the year.

CALENDAR
Sow – early to late spring.
Harvest – year round.

SOIL AND SITE
- Beet leaf has the same needs as spinach – it likes well-manured, deeply dug soil in a sunny position.
- Beet leaf can be grown on poorer soil, but does best on a rich, well-drained, moist soil.
- Some growers favor planting the crop in raised beds.

SOWING AND PLANTING
- Sow seeds in early to late spring in ¾ in. deep drills set 15 in. apart, with a pinch of 3–4 seeds at 9 in. intervals.
- Compact the soil and water generously.
- When the seedlings are big enough to handle, pinch out to leave the strongest plants.
- Firm up the soil around the plants.

PLANT CARE
Stir the soil with a hoe to create a loose-soil mulch. In dry weather, spread a mulch of old manure to hold in the moisture.

HARVESTING
Harvest from mid-summer round the year – the dates depend on the variety, where you live, and how much protection you give the plants. Pick young leaves close to the ground. Remove old and yellow leaves.

VARIETIES
Rhubarb Chard: Looks a bit like young rhubarb. Everything can be eaten – the leaves cooked like spinach, and the stems steamed like asparagus.
Swiss Chard: Sometimes called Beet Chard, and Seakale Beet, it produces dark, glossy leaves on long, white, celery-like stems.
Yellow Chard: Well-known, easy-to-grow variety that produces bright green, fleshy leaves on yellow stems.

BEETROOT

FACTS/TIPS
- Beetroots don't like to be disturbed, so be careful when weeding and thinning.
- For maximum sweetness, cook straight from picking.
- Wash away the soil; otherwise they taste dull and muddy.
- The 'bottom line' is that lazy bowels love beetroot!

CALENDAR
Sow – early to mid-spring; early to mid-summer.
Harvest – late spring to early summer; mid to late autumn

SOIL AND SITE
- Beetroot prefers a light sandy soil that has been manured for a preceding crop.
- Traditionally, growers on clay soil sow a couple of weeks later than usual so as to keep the roots small and compact.
- The ground needs to be well drained and manured.

SOWING AND PLANTING
- Sow seeds in early to mid-spring in 1 in deep drills 10–12 in. apart. Sow a pinch of 3–4 ready-soaked seeds at 5 in. intervals.

- Compact the soil and water generously.
- Pinch out the seedlings to leave the strongest plants.
- Firm the soil up around the stems of the remaining plants, and water generously with a fine spray.

PLANT CARE

Spread a web of black cotton thread over the young plants to keep off birds. Thin out to 8 in. apart and use the baby beet for salad. Stir the soil with a hoe to create a loose-soil mulch. In dry weather, spread a mulch of spent manure to hold in the moisture. Water frequently.

HARVESTING

Harvest from late spring through to late autumn, the date depending on the variety. Use a fork to ease the root from the ground. Twist the leaves off about 2 in. above the crown and eat fresh, or store in sand.

VARIETIES

Detroit: Traditional variety that produces good, solid roots.
Mammoth Long: Good, reliable, easy-to-grow variety that produces long, smooth, cylindrical roots.
Tardel: Late-sowing variety that produces sweet, globe-shaped, baby beetroots.

BROCCOLI

FACTS/TIPS

- Homegrown broccoli is many times better than shop-bought.
- You can extend the cropping time by picking early at the shooting stage and picking late when the little flowers appear.

CALENDAR

Sow – early to late spring
Plant – early to late summer
Harvest – depending on the varieties, mid to late winter, mid to late spring, and mid-summer to mid-autumn.

SOIL AND SITE

- Can be grown on a sandy soil, but prefers a heavy, fertile loam that is inclined to clay.

- The soil should be well manured and compact, and deeply dug for a previous crop.
- Compact soil produces compact heads on the broccoli.
- Plant in an open, sunny, sheltered spot.

SOWING AND PLANTING

- Sow seeds in mid- to late spring in prepared seedbed or trays.
- Plant out in early to mid-summer on a dull, rainy day.
- Dib (dibble) holes 18–27 in. apart, with 18–27 in. between rows, 'puddle' the seedlings into the holes and compact the soil.

PLANT CARE

Water the seedlings before and after planting. Stir the surface soil with a hoe to create a loose-soil mulch. In dry weather, spread a mulch of spent manure to hold in the moisture.

HARVESTING

Harvest from mid-winter to late spring, and mid-summer to late autumn, depending on the variety and how much protection you give the plants. Start by picking the central spears and follow up by picking the little sideshoots. Pick every few days to encourage new growth.

VARIETIES

Purple Sprouting Early: Sprouting variety that produces a wealth of dark purple spears in early spring.
Purple Sprouting Late: Much the same as the early purple, but can be cropped from mid-spring.
Red Arrow: Variety that produces an abundance of bright purple flowers or spears.

BRUSSELS SPROUTS

FACTS/TIPS

- Go for tight, firm varieties and steam rather than boil.
- Brussels prefer a firm soil in a sunny, weed-free situation.
- Choose standard old varieties like 'Bedford' if you like large strong-tasting sprouts, and F1 hybrids like 'Igor' and 'Oliver' for sweet ones.

CALENDAR

Sow – early to late spring.
Plant – late spring to early autumn.
Harvest – First crop mid to late winter or late winter to early spring. Second crop early autumn to early winter.

SOIL AND SITE

- Brussels prefer a deeply worked, rich, fertile, firm soil.
- A good option is to manure and dig for a preceding crop and then to plant out the sprouts.
- The plot needs to be sunny and open, but free from winds.
- If your plot is windy, go for low-growing varieties.

SOWING AND PLANTING

- Sow seeds in early to mid-spring in a prepared seedbed or trays.
- Plant out in late spring to early summer on a dull, rainy day.
- Firm the ground. Dib (dibble) holes 20–36 in. apart, with 20–36 in. between rows, and 'puddle' the seedlings into the holes.

PLANT CARE

Water the seedlings both before and after planting. Stir the surface of the soil with a hoe to create a loose-soil mulch. Remove the bottom leaves as they become yellow. Spread a mulch of old spent manure or straw over the loose soil to hold in the moisture.

HARVESTING

Harvest from early autumn to early spring, the dates depending on the variety and growing methods. Work from bottom to top up the stem picking the best tight sprouts.

VARIETIES

Evesham Special: Dependable traditional English variety that is very prolific.
F1 Oliver: Early disease-resistant variety that crops from late summer through to mid-autumn.
F1 Igor: Vigorous, frost-resistant, high-yielding variety that produces masses of tight, round sprouts.

CABBAGES

FACTS/TIPS
- You can eat cabbage all year round.
- Plant spring varieties every 4 in. and eat immature plants as 'spring greens.'
- Cabbage is best lightly steamed for 5–8 minutes.

CALENDAR
SPRING CABBAGE
Sow – early to mid-summer.
Plant – early to mid-autumn.
Harvest – late winter to early summer.
SUMMER CABBAGE
Sow – late winter to late spring.
Plant – mid-spring to mid-summer.
Harvest – mid-summer to mid-autumn.
WINTER CABBAGE
Sow – mid to late spring.
Plant – early to mid-summer.
Harvest – early to mid-summer and mid-autumn to early winter.

SOIL AND SITE
- Prefers a deeply worked, firm, well-drained, moist, well-manured soil on a sheltered, wind-free site.
- The soil must be firm and compact.

SOWING AND PLANTING
- Sow seeds – spring varieties in mid to late summer; summer varieties in late winter to late spring; winter varieties in early to late spring – sow seeds in a prepared seedbed or in seed trays (flats).
- Plant out – spring varieties in early to mid-autumn; summer varieties in mid-spring to early summer; winter varieties in early to mid-summer.
- Dib (dibble) holes 12–14 in. apart, with 12–14 in. between rows, and 'puddle' the seedlings into the holes.

PLANT CARE
Water daily. Use the hoe to pull the soil up around the plants. In very dry conditions, spread a deep mulch of old spent manure or straw over the loose soil to hold in the residual moisture.

HARVESTING
Harvest from mid-winter to the following autumn, the precise time depending on your chosen varieties and growing methods. Use a knife to cut the cabbage off close to ground level and slash a deep 'X' on the stump to encourage a crop of secondary mini-cabbages.

VARIETIES
F1 Tundra: Swift-growing, frost-hardy, winter-harvesting variety.
Pointed Durham Early: Spring-harvesting, pointed-head variety that can be cut as spring greens.
Spring Hero F1: Spring-harvesting variety that produces a huge, solid, medium-green head with a white heart.

CAPSICUMS (SWEET PEPPERS)

FACTS/TIPS
- You can sow capsicums directly in growing-bags, or sow them in trays and then plant out.
- A good option is to set a wigwam of bamboo canes over the plant and wrap it around with plastic sheeting.

CALENDAR
Sow – late winter to early spring.
Plant – mid-spring to early summer.
Harvest – mid-summer to mid-autumn.

SOIL AND SITE
- Capsicums prefer a well-drained, well-manured soil in a sunny, sheltered position.
- A good option is to plant them in a raised bed with a plastic-screen windbreak all around.
- You could plant them in growing-bags in a small greenhouse or in a pot with some sort of dedicated glass or plastic shelter.

SOWING AND PLANTING
- Late winter or early spring – sow the seeds in trays on a bed of moistened potting compost, and protect with a sheet of glass topped with newspaper.
- Mid to late spring – prick the seedlings out into 3 in. peat pots, water and keep warm.
- Plant outside in early summer. When the plants are strong enough, pot on into 9 in. pots, and protect with a cloche.

PLANT CARE
Water the seedlings before and after planting. Stir the surface of the soil with a hoe to create a loose-soil mulch. Support the plants with a cane. Spread a mulch of old spent manure or straw over the loose soil to hold in the moisture.

HARVESTING
Harvest from mid-summer to mid-autumn, the date depending on the variety. Use a sharp knife to cut the fruits as needed.

VARIETIES
D'Asti Giallo: Medium-early variety that produces brilliant yellow, thick-fleshed fruits that turn from green to red on maturity.
F1 Beauty Bell: Sweet variety that produces an abundance of large, chunky fruits.
Sweet Spanish Mixed: Selection of varieties that gives a mixture of colors and shapes.

CARROTS

FACTS/TIPS
- The British eat more carrots than anyone else.
- You can be eating carrots for the best part of the year.
- Growing carrots alongside onions cuts down on carrot fly.

CALENDAR
Sow – early spring to late summer.
Harvest – early summer to early winter.

SOIL AND SITE
- Carrots prefer a friable, deeply dug, well-drained, sandy, fertile loam in a sunny position.
- If the soil is heavier, then choose short, stumpy varieties.
- If your soil is stony, the growing root will probably stunt or fork.

SOWING AND PLANTING

- Sow seeds in early spring through to early summer depending upon variety. Run ¾ in. deep drills 6–9 in. apart, sow thinly and water with a fine sprinkler.
- Thin the seedling to 2 in. apart.
- Shelter the plants at the beginning and end of the season.

PLANT CARE

Hoe frequently to create a loose-soil mulch to help retain the moisture. Be careful not to damage the tender plants.

HARVESTING

Harvest from late spring through to early winter. Make successive sowings of a range of varieties. Store maincrop carrots in boxes of sand in a frost-free shed.

VARIETIES

Chantenay Red Cored: Old, tried-and-trusted, mid-season or maincrop variety that produces stump-ended roots.
Early Nantes: Very reliable, traditional, mid-season variety that produces long, blunt-ended roots.
F1 Flyaway: The first fly-resistant variety; produces large, stumpy-shaped, smooth-skinned carrots.

CAULIFLOWERS

FACTS/TIPS

- Experienced growers reckon that if you can grow a cauliflower then you can grow anything.
- Cauliflowers need lots of water.
- They are very sensitive to poor soil, or extremes of weather – too dry, too wet, or too cold.

CALENDAR

Sow – early to late spring.
Plant – early to mid-summer.
Harvest – depending upon variety, for most of the year.

SOIL AND SITE

- Cauliflowers prefer a deep, well-manured, compact, well-drained, moisture-retaining soil in sun.

- The ideal is to manure for some other crop, and then follow on with cauliflowers.
- If the soil is really poor and/or too dry, or in any way grossly undernourished, the heads will come to nothing and you will be disappointed.

SOWING AND PLANTING

- Sow seeds in early to late spring in prepared seedbeds or trays.
- Plant out in early to mid-summer, so that you don't have to water.
- Dib (dibble) holes 20–24 in. apart, with 20–24 in. between rows, and 'puddle' the seedlings into the holes.
- Firm up the soil around the plants and water generously.
- Fit loose-fit plastic/felt collars to protect from root fly.

PLANT CARE

Water the seedlings before and after planting. Stir the surface of the soil with a hoe to create a loose-soil mulch, and spread a mulch of chopped straw to further hold in the moisture. Water copiously.

HARVESTING

If you sow early, autumn and winter varieties, you can harvest from one summer around to the next. Cut complete with the wrap-around leaves, or lift the whole plant and store in a frost-free shed.

VARIETIES

Autumn Giant: Old reliable summer and autumn harvesting variety that produces large, solid, white heads.
Dominant: Good variety for summer and autumn harvesting; produces large, compact, slightly creamy-colored heads.
White Rock: Good variety for harvesting from late summer onwards; produces beautiful, tight, white heads well hidden away in a shell of leaves.

CELERIAC

FACTS/TIPS

- Celeriac must be well watered, sown under glass, and carefully hardened off before planting out.
- A celery heart and a piece of young celeriac taste similar, but that is about it.
- Tastes best when cooked like rutabagas, or eaten raw with a salad.

CALENDAR

Sow – early to mid-spring.
Plant – late spring to early summer.
Harvest – early autumn around to late spring the following year, and longer if you provide protection.

SOIL AND SITE

- Celeriac prefers a well-manured soil, with the manure having been dug in the preceding winter, on an open, sunny but sheltered site.
- Needs unlimited water.
- Young plants don't like cold winds.

SOWING AND PLANTING

- Sow seeds in early to mid-spring under glass in prepared seed trays.
- Pot the seedlings on as soon as they are big enough to handle.
- Plant out in late spring to early summer 1 ft apart, with 10–15 in. between rows. Set the plants as 'shallow' as possible.

PLANT CARE

Water daily. Hoe to create a loose-soil mulch. Remove old leaves and any wandering shoots and roots. Never allow the ground to dry out.

HARVESTING

If your site is well drained, and you protect the plants with straw and fleece, you can leave the crop in the ground, and harvest from late autumn around to the following early spring. If your soil is wet, then lift and store in a frost-free shed.

VARIETIES

Balder: Well-tried variety that produces large, round, brown roots with lots of dark green foliage.
Giant Prague: Reliable variety that produces large roots that can be harvested from early autumn onwards.

Prinz: Variety that resists leaf disease and bolting. Has a thin, light skin and crisp, white flesh; good with a wedge of strong cheddar cheese.

CELERY

FACTS/TIPS
- Growing celery is very time-consuming.
- Self-blanching varieties negate the need for earthing up.
- If you grow a mix of traditional and self-blanching varieties, you can be eating celery from mid-autumn through to late winter.

CALENDAR
Sow – early to mid-spring.
Plant – late spring to early summer.
Harvest – late summer to late winter.

SOIL AND SITE
- Celery does best in a deep, rich, heavy, moist, well-drained soil in a sunny position.
- Celery needs lots of moisture but will not thrive if the soil is waterlogged or sour.
- Prepare the ground with well-rotted manure. The piled earth – for earthing up – must be fine and friable.

SOWING AND PLANTING
- Sow seeds in early to mid-spring under glass in prepared seed trays.
- Pot the seedlings on, as soon as they are big enough to handle, into peat pots.
- Plant out in late spring to early summer; dig a trench 1 ft deep and 22 in. wide. Put manure topped with earth into the trench, and set the plants 9 in. apart.

PLANT CARE
Water the plants daily. Stir the surface of the soil with a hoe to create a loose-soil mulch. Remove suckers and dead leaves. When the plants are about 10–12 in. high, bind them with raffia, and then repeatedly earth up so as to create a ridge or mound that covers all but the foliage. Cover with chopped straw in frosty weather.

HARVESTING
Harvest self-blanching varieties from late summer through to late autumn, and traditional varieties from late autumn around to the following early spring. Use a fork to ease the plant from the ground.

VARIETIES
Giant Red: Hardy variety with a slightly loose and open form; produces very solid, dark red-purple heads that blanch to pink-white.
Giant White: Popular variety much like Giant Red.
Pascal: A reliable trench variety that produces solid heads of crisp white-to-green stalks.

CHICORY

FACTS/TIPS
- The bitter 'smoky burnt' taste of chicory nicely offsets sweeter foods like bread, butter, and cheese and is perfect in a sandwich.
- A challenging and fun aspect of growing chicory is that you can take the stored roots three or four at a time – from late autumn to the following spring – and grow them up as fat buds or chicons.
- The lettuce-like chicory varieties are easier to grow but less interesting.

CALENDAR
Sow – depending upon variety, late spring to early summer.
Harvest – mid-winter to mid-spring and mid-autumn to early winter.

SOIL AND SITE
- Prefers a light to medium, soft-textured, deeply dug, fertile, moist soil.
- Avoid freshly manured ground, because the roots of the growing chicory will hit the fresh manure and divide.
- The ideal is to spread the manure for one crop and then follow on with the chicory.
- Choose a sunny corner, well away from draughts.

SOWING AND PLANTING
- Sow seeds in late spring to mid-summer in ½ in. deep drills 10–12 in. apart.
- Water with a fine sprinkler.
- Thin the seedlings to 8 in. apart.

PLANT CARE
Water daily. Stir the surface of the soil with a hoe to create a loose-soil mulch. When the leaves have died down, lift and trim the roots to remove forked roots and earth and bed them down flat in dry sand.

HARVESTING
Harvest and eat self-blanching varieties from mid-autumn to early winter. You can blanch traditional types from late autumn to mid-spring. Take four roots at a time – ones that you stored in autumn – plant them in damp sand and cover with black plastic. They will be ready in 3–4 weeks.

VARIETIES
Brussels Witloof: Reliable, easy-to-grow variety that produces good-sized, carrot-like roots and medium-sized, yellow-white chicons.
F1 Zoom: Easy-to-blanch variety that produces carrot-like roots from early autumn onwards.
Red (Radicchio) Variegata di Chioggia: Variety that produces large multi-colored ball or rosette-like chicons.

CUCUMBERS

FACTS/TIPS
- There are indoor and outdoor varieties.
- Choose a variety to suit your needs; for example, if you are growing them outdoors, make sure you choose ridge or gherkin types.
- Growing-bags are a good option – sow the seeds directly in the bags.

CALENDAR
OUTDOOR
Sow – late spring.
Plant – early summer.
Harvest – mid-summer to mid-autumn.

INDOOR
Sow indoors – mid-spring.
Plant out – late spring.
Harvest – early summer to early autumn.

SOIL AND SITE

- Cucumbers prefer a light to medium, soft-textured, deeply dug, well-manured, moist soil.
- For outdoors, dig pits or a trench – depending upon how many plants you have in mind to grow – and top up with a mix of compost and/or well-rotted manure.
- Choose a spot in full sun, well away from draughts, and then put a plastic screen or shelter on the windward side.

SOWING AND PLANTING

- For indoor varieties, in mid-spring sow the seeds in trays on moistened tissue. Keep warm.
- In late spring, prick the seedlings out into growing-bags and cover with glass/plastic sheeting.
- For outdoor varieties, in late spring or early summer dig a trench 1 ft deep and 1 ft wide, and fill with manure and soil. Sow three seeds ¾ in deep at 2–3 ft intervals and cover with plastic.
- Thin to the best plant. Cover with glass/plastic sheeting.

PLANT CARE

Support indoor plants with canes and pinch out the growing tip when it reaches the top of your greenhouse or cloche. For outdoor plants, let the plant trail along the ground like a zucchini. Stop the side-growing shoots when they get to your frame/ boundary. Water daily.

HARVESTING

Harvest indoor varieties from mid-summer to mid-autumn, and outdoor varieties from early summer to early autumn. Support the weight of the fruit and cut them off with a sharp knife.

VARIETIES

Bedfordshire Prize: Outdoor variety that produces an abundance of short, slightly ridged fruits.
King George: Indoor variety with long, straight, dark green fruits.

Telegraph Improved: Improved traditional variety that produces good, straight fruits.

ENDIVE

FACTS/TIPS

- Endive is like lettuce – only more interesting.
- It needs to be blanched.
- Blanching takes 2–3 weeks; in that time, the taste shifts from being bitter through to being sweet with a slightly tart background tang.
- Some varieties have such large, bushy heads that they are, in effect, self-blanching.

CALENDAR

Sow – early spring to early autumn.
Harvest – late summer around to mid-spring.

SOIL AND SITE

- Endive prefers a deeply dug, rich, medium light, fertile soil that contains plenty of well-rotted manure from a previous crop.
- It needs a sunny open spot for summer and autumn varieties, and partial shade for spring-grown ones.

SOWING AND PLANTING

- Sow seeds in early spring to early autumn in ½ in. deep drills 10–12 in. apart. Sow thinly and water generously.
- Thin the seeds to 1 ft apart.
- Water generously in dry weather.

PLANT CARE

Stir the soil with a hoe to create a loose-soil mulch. Endive must be blanched – it is useless otherwise. Blanch the plants when they are fully grown. On a sunny day when the air is dry, gather the leaves together, bind them with raffia, and cover with a clay flowerpot.

HARVESTING

Harvest 2–3 weeks after covering the plant with the flowerpot. Check the plant every day or so to make sure it is dry and free from slugs. After two weeks, lift the pot and use a knife to cut the plant off close to the ground.

VARIETIES

Pancalieri: Popular variety that produces heads of dark green, lace-edged leaves.
Scarola Verde: Swift-blanching variety for spring and summer use that produces very large, frilly-edged, green-white heads.
Wallone: Self-blanching French type that produces a large, densely packed heart.

KALE

FACTS/TIPS

- Kale is very easy to grow.
- A good option if you have trouble with club root, heavy frosts, poor soil and so on.
- It can be cropped from mid-autumn right around to early spring.
- Hardy low-growing varieties are good for windy sites.

CALENDAR

Sow – mid-spring to early summer.
Plant – early to late summer.
Harvest – late autumn to late spring.

SOIL AND SITE

- Kale prefers a strong, deeply worked, very well-compacted, fertile loam.
- A good option is to spread the manure for one crop, and then follow on with the kale.
- Kale will put up with just about anything, apart from loose ground, standing water or a long, hard, bitter frost.

SOWING AND PLANTING

- Sow seeds in mid to late spring in prepared seed beds or trays.
- Plant out in early to late summer on a dull rainy day.
- Dib (dibble) holes 15–18 in. apart, with 18–20 in. between rows, and 'puddle' the seedlings into the holes.
- Water frequently.

PLANT CARE

Water daily. Stir the surface of the soil with a hoe to create a loose-soil mulch, and draw the earth to support and to protect from wind and frost. If it is very windy, rig up some sort of screen on the windward side.

HARVESTING

Harvest from late autumn to late spring. Use a knife to nip out the crown, and then work from top to bottom picking off the sideshoots. Remove all the yellow leaves.

VARIETIES

Cottagers: Traditional hardy variety that produces an abundance of medium-green leaves.

Dwarf Green Curled: Old variety that produces a mass of densely curled, dark green leaves on a central stalk.

Red Russian: Large variety that produces lots of crinkly, green-purple leaves on a tall stem.

KOHL RABI

FACTS/TIPS

- Kohl rabi is delicious when eaten raw in a salad, and also when steamed and mashed like a rutabaga.
- It is best eaten when it is young and tender.
- A variety called Superschmelz can grow to a gigantic nearly 18 lb!

CALENDAR

Sow – early spring to late summer.
Harvest – early summer to early winter.

SOIL AND SITE

- Kohl rabi prefers a light to medium, soft-textured, moist, fertile soil.
- It prefers a sheltered spot in dappled shade, and plenty of water.
- The plants are prone to bolting in dry weather.

SOWING AND PLANTING

- Sow seeds in early spring to late summer, depending upon variety, in ¾ in. deep drills 10–12 in. apart, a pinch of seeds every 5–6 in.
- When the seedlings are big enough to handle, thin out to leave the strong plants at 6 in. intervals.
- Water generously in dry weather.

PLANT CARE

Hoe to create a loose-soil mulch and water generously. In dry weather, spread a mulch of spent manure or chopped straw. Remove yellow leaves. Remove sideshoots and foliage so as to keep a smooth, rounded shape.

HARVESTING

Harvest from early summer to early winter, the precise time depending on the variety. Use a fork to ease the globe out of the ground. Lift as needed. Kohl rabi spoils if left in the ground too long, and if stored.

VARIETIES

F1 White Danube: Popular late variety that produces a white-fleshed, green-skinned root.

Green Delicacy: Early variety that produces a fist-sized root with a pale green skin and a white flesh.

Purple Delicacy: Late, hard variety that produces a root with a purple skin and white flesh.

LEEKS

FACTS/TIPS

- Leeks need to be earthed up and blanched, just like celery.
- They are good for winter use, when vegetables are scarce.
- Varieties range from small and mild through to large and strong-tasting.

CALENDAR

Sow – early to late spring.
Plant – early to late summer.
Harvest – early autumn around to late spring.

SOIL AND SITE

- Leeks prefer a deeply worked, well-manured, moist soil.
- A good method is to dig holes or trenches and set the leeks in a nice mix of old, well-rotted manure and compost, with piled earth to the side in readiness for earthing up.

SOWING AND PLANTING

- Early crops – sow in mid to late winter in trays under glass/plastic. Prick out to 2 in. apart, and plant out in mid-spring.

- Main crops – sow in early to mid-spring in ½ in. deep drills 10–12 in. apart. Sow thinly and water with a fine spray. Thin out to a spacing of 1 in.
- Plant in late spring to early summer on a wet, showery day. Dib (dibble) 6 in. deep holes at 6 in. intervals, with 1 ft between rows. Transplant the seedlings, one plant per hole, and top up with water.

PLANT CARE

Hoe to create a loose-soil mulch. Ring the established seedlings around with a mulch of chopped straw or old spent manure. Throughout the season, draw the earth up over the plants until all but the tops are covered. Water frequently.

HARVESTING

Harvest from early autumn to late spring. Use a fork to ease the roots from the ground and lift as needed.

VARIETIES

Giant Winter: Late variety that produces heavy, thick stems from mid-winter onwards.

Musselburgh: Traditional, winter-hardy variety that produces good-sized stems from early winter onwards.

Prizetaker/Lyon: Reliable early-autumn variety that produces long, thick stems.

LETTUCES

FACTS/TIPS

- You can grow lettuces right through the year.
- Lettuces intercropped with radishes produce a good salad all in one go.
- For best flavor and texture, take a fresh-picked, organic, home-grown lettuce, remove the outer leaves and eat the heart as it is.
- Most children like lettuce hearts dipped in a dressing and wrapped around with fresh bread.

CALENDAR

SUMMER to EARLY WINTER
Sow – early spring to early summer.

Harvest – late spring to mid-autumn.
SPRING
Sow – late summer to mid-autumn.
Harvest – mid to late spring.

SOIL AND SITE

- Lettuce prefers a light, friable, well-manured, well-drained, moisture-retentive, sandy loam.
- A good growing plan is to build raised beds with boards all around, and to fill the bed with a mix of sandy loam, compost and old, well-rotted manure.

SOWING AND PLANTING

- Sow thinly in ½ in. deep drills 10–12 in. apart. Compact the soil and spray with water. Protect with glass/plastic and thin to 1 ft apart.
- Thin to 9–10 in. apart so that lettuces are in staggered rows.

PLANT CARE

Hoe to create a loose-soil mulch. Run a web of black cotton threads over the plants to keep off the birds. Water generously in dry weather.

HARVESTING

Harvest from mid-spring to mid-autumn, depending on your chosen varieties. If you protect the crop with fleece/plastic you will be able to sow early and crop late. Cut close to the ground as needed.

VARIETIES

All Year Round: Hardy, traditional, spring- and summer-sowing Butterhead variety that produces medium-sized compact hearts with a crisp texture and a sweet-tangy flavor.
Little Gem: Small Cos-like variety that produces small, compact heads.
Webb's Wonderful: Reliable, slow-bolting variety that produces frilly, crisp-textured heads with a delicate, sweet taste.

MARROWS AND COURGETTES (ZUCCHINI)

FACTS/TIPS

- Stuffed and roasted, or steamed, marrows are good; fried courgettes are better.
- Marrows are good if you like big, but courgettes are easier to grow and tastier.

CALENDAR

Sow – mid to late spring.
Plant – late spring to early summer.
Harvest – mid-summer to mid-autumn.

SOIL AND SITE

- Marrows and courgettes prefer deeply worked, well-manured, moist but well-drained soil.
- Dig a hole/trench, and fill it with manure topped with a mound of soil.
- Set the plants in place in the mound.
- The mound catches the sun, and ensures that the soil is well drained.
- Once the plants are hardened and under way, cover the mound with a mulch to hold in the moisture.

SOWING AND PLANTING

- Sow seeds in mid- to late spring with 2–3 seeds in peat pots under glass/plastic. Thin to one good plant.
- Plant in late spring to early summer. Dig holes 1 ft deep, 1 ft wide, in lines at a spacing of 4–5 ft apart, fill with well-rotted manure and mound over with soil. Set the peat pot plants into the top of the mound and water generously.

PLANT CARE

Hoe to create a loose-soil mulch. Spread a mulch of spent manure around the plant. Pinch the tips out at about 2 ft. Keep watering and mulching throughout the season with grass clippings, more spent manure.

HARVESTING

Harvest from July through to October. Cut courgettes every few days to keep them producing. Cut marrows as needed. At the end of the season, marrows can be hung in nets and stored in a frost-free shed.

VARIETIES

All Green Bush: High-yielding courgette variety that produces masses of dark green fruits.
Long Green Bush: Prolific marrow variety that produces large fruits.
Zucchini: Courgette variety that produces small, dark-skinned, tight-fleshed fruit.

ONIONS AND SHALLOTS

FACTS/TIPS

- If you like onions in soups, salads, sandwiches, and so on, you need to grow a whole range of varieties.
- Onions can be grown from seed or sets; sets are swifter, but more expensive.
- If you grow a mix of onions and shallots, you can be eating them from early summer to mid-autumn.

CALENDAR

ONIONS FROM SETS
Sow – late winter to late spring.
Harvest – early summer to early autumn.
ONIONS FROM SEED TO EAT IN SUMMER
Sow – late winter to mid-spring.
Harvest – early summer to mid-autumn.
ONIONS FROM SEED TO OVERWINTER
Sow – late summer to mid-autumn.
Harvest – early summer to early autumn.

SOIL AND SITE

- Onions prefer a deeply worked, well-manured, moist, well-drained, friable, fertile, sandy soil in a sunny position.
- Ideally, the manure needs to be dug in during the autumn so that it is ready for spring sowing.
- Some enthusiasts reckon that you cannot do better than spread a topdressing of dried chicken manure and sand.
- Sandy soil needs a dressing of ground clay; clay soil needs a dressing of sand or grit.

SOWING AND PLANTING

- Spring-sown seed – in late winter to mid-spring sow thinly in ½ in. deep drills 10 in. apart; thin to 1–2 in. apart, depending upon variety.
- Summer-sown seed – in late summer to early autumn sow thinly in ½ in. deep drills 10 in. apart; thin to 2 in. apart.
- Spring sets – in late winter to mid-spring plant 2–3 in. apart, in rows 10 in. apart.
- Exhibition onions – in mid-winter sow in a tray and keep on a windowsill. Plant out in mid-spring 3 in. apart, with rows 12–15 in. apart.

PLANT CARE

Hoe to create a loose-soil mulch. Later, use your fingers to draw the soil slightly away so that the swelling bulb sits high on the surface. Don't water, but keep stirring the ground to prevent the moisture wicking out from the underlying moist soil.

HARVESTING

Harvest from early summer to early autumn. Lift salad onions, and large onions, as needed. To ripen and store, first bend the tops over. When the tops are yellow, lift the onions and lay them in the sun to dry. Finally, put them in boxes, or string them up, and store in a frost-free shed.

VARIETIES

Alisa Craig: Old-fashioned, tried, tested and tasted exhibition variety that produces large, golden-skinned, mild-flavored onions.
Bedford Champion: Very popular, old-fashioned variety that produces large, tasty onions that store well.
Senshyu Yellow: Overwintering, heavy-yielding, Japanese variety that produces good-sized, globe-shaped, white-fleshed onions.

PARSNIPS

FACTS/TIPS

- Parsnips are hardy and easy to grow.
- They take quite a time to germinate, so be patient.

- Most of us know about the joys of roast parsnips, but they are just as good boiled and mashed, or steamed, covered in grated cheese and then grilled.
- Parsnips can be left in the ground over winter and lifted and eaten from mid-autumn to the following spring.

CALENDAR

Sow – late winter to mid-spring.
Harvest – mid-autumn to early spring.

SOIL AND SITE

- Parsnips prefer a deeply dug, friable, fine-textured, well-drained, fertile soil.
- Parsnips prefer soil that has been well manured for a preceding crop.
- Be warned: if you use fresh manure, the growing roots will either fork or become cankered.
- Short, stubby varieties are good for stony ground.

SOWING AND PLANTING

- Sow seeds in late winter to mid-spring in ¾ in. deep drills 8–10 in. apart, a pinch of 3–4 seeds at 5–6 in. intervals.
- Compact the soil and water generously with a fine spray.
- When the seedlings are big enough to handle, pinch out to leave the strongest plant.

PLANT CARE

Hoe to create a loose-soil mulch. Be very careful that you don't scuff or graze the emerging root; such damage often results in top-rot or canker. Water little and often so as to avoid wet-drought-wet conditions that result in root splitting.

HARVESTING

Harvest from mid-autumn to early spring. Parsnips are frost-hardy, so you can leave them in the ground until they are needed. When you come to lifting, ease them up with a fork, so that you don't break off the long root tips.

VARIETIES

F1 Countess: Disease-tolerant, high-yielding maincrop variety that produces smooth-skinned roots.

Hollow Crown: Reliable old variety for a good deep soil that produces long, tapered, white-skinned, creamy-fleshed roots.
White King: Classic long-rooting variety for deep soil that produces tapered, medium-sized, creamy-skinned, white-fleshed roots.

PEAS

FACTS/TIPS

- Peas are a pleasure to grow, relaxing to shuck, and sweet to eat.
- Kids enjoy eating freshly picked peas straight from the pods.
- If you sow in succession – first early, second early and maincrop – you can be eating freshly picked peas from spring to mid-autumn.
- With varieties like Mangetout and Sugar Snap, you can eat pods and all.

CALENDAR

Sow – depending upon variety, early spring to mid-summer.
Harvest – late spring to mid-autumn.

SOIL AND SITE

- Early crops do best on warm, dry, sandy soil, while maincrops prefer a heavier, richer, moisture-retentive loam.
- Be warned: if the manure is fresh, the peas will roar out of control to become all leaf and few pods.
- Peas need lots of moisture. If the plant starts to dry out, soak it with water and cover it with a mulch of old, well-rotted, spent manure.

SOWING AND PLANTING

- Sow seeds in succession from early spring to mid-summer in 2 in. deep drills, in rows 15–48 in. apart, depending upon height of variety. Sow seeds at 5–6 in. intervals.
- Compact the soil and water generously.
- Cover the rows with something like wire mesh, twigs, cottons – anything to keep the birds away.

CROP CARE

Hoe the soil to create a loose-soil mulch, especially in dry weather. In long dry spells, drench the soil with water and put a heaped line of old manure mulch on each side of the row.

HARVESTING

Harvest from early summer to mid-autumn. Gather the pods when they are still young, every 2–3 days, to encourage new pods to develop.

VARIETIES

Alderman: Traditional, tall-growing, maincrop variety, growing to 3–5 ft, that produces masses of large pods.
Avola: Reliable, low-growing, heavy-cropping, first-early variety that produces good-sized, sweet, tasty peas.
Greenshaft: Very popular, low-growing, heavy-cropping, second-early variety, growing to about 2 ft high, that produces masses of long, pointed pods in pairs.

POTATOES

FACTS/TIPS

• Many people eat potatoes every day.
• Potatoes are easy to grow, easy to keep, and tasty.

CALENDAR

Planting – early to late spring.
Harvest – early summer to mid-autumn.

SOIL AND SITE

• A wet soil tends to result in a slick, soapy, slightly yellow potato, while potatoes grown on dry, sandy soil are often loose and fluffy.
• A low-lying soil will produce good potatoes in a dry year, but there is the potential for disease if the weather is humid.
• The soil needs to be deeply dug, well-drained, friable, clay to sandy, and moisture-retentive, in a sunny position.

SOWING AND PLANTING

• Chitting for early crop – in late winter sit the seed potatoes in trays in a shed, until there are 1 in. long shoots.
• Sowing early – in early to late spring, in 6 in. deep drills, set the chitted seeds 12–16 in. apart, in rows 24–30 in. apart. Cover the shoots.
• Sowing under plastic – in early to late spring rake and water the soil. Set the chitted seeds on the surface, 12–16 in. apart, in rows 24–30 in. apart. Cover with a long mound of soil topped with black plastic sheet. Cut slits at each planting point.

PLANT CARE

As soon as the foliage shows, hoe to form a ridge. Repeat this procedure so that the foliage is always just showing at the top of the ridge. In dry weather, spread a mulch of spent manure or chopped straw around the plants.

HARVESTING

Harvest from early spring to mid-autumn. Use a flat-pronged fork to work from the outer limits of plant towards the center. Leave the spuds on the surface to dry and then bag them up into 'good' and 'damaged.' Eat the damaged ones first. Lift new potatoes as needed. Store surplus maincrop potatoes in shallow boxes in a dry, dark, frost-free shed, or in a clamp. To build a clamp, heap the dry potatoes and cover the heap with straw topped with earth.

VARIETIES

Desiree: Drought-resistant, early main variety that produces a medium-sized, red-skinned, cream-fleshed tuber.
King Edward: Popular, early main variety that produces a pink/cream-skinned, creamy-fleshed tuber.
Marfona: Second-early variety that produces large, smooth-skinned tubers.

RADISHES

FACTS/TIPS

• Radishes can easily be grown all year round.
• The secret of growing a swift, plump crop is to choose the appropriate variety, and to sow them in rich, moist soil.
• You can plant radishes as a swift catch crop between crops such as cauliflowers, peas and lettuces.

CALENDAR

Sow – mid-winter to late summer.
Harvest – mid-spring to late winter.

SOIL AND SITE

• Radishes prefer a rich, moist, fertile soil in a sunny position – a soil that is rich in humus from a previous crop.
• Radishes do well on soil made up from semi-exhausted manure-peat compost gathered from hot beds and growing-bags.
• Be warned: fresh manure will result in fast-growing, leafy plants that are tough and stringy at the root.

SOWING AND PLANTING

• Sow seeds every few days in succession from mid-winter to late summer in ½ in. deep drills 4–6 in. apart.
• When the seedlings are big enough to handle, thin to leave the strongest plants 1 in. apart. Water generously before and after thinning.

PLANT CARE

Hoe at each side of the row to create a loose-soil mulch. Water little and often to avoid the wet-drought-wet conditions that result in root splitting. In dry weather, hoe a mulch of spent manure up against each side of the row.

HARVESTING

Harvest from mid-spring to late winter. Pull the radishes when they are young and tender. Eat them as soon as possible after pulling.

VARIETIES
Cherry Belle: Mild-flavored spring and summer variety that produces bright red, cherry-like, white-fleshed roots.
French Breakfast: Very popular spring and summer variety that produces a long, red, white-tipped root.
Scarlet Globe: Popular, quick-growing, early-cropping, tried, trusted and much tasted, traditional variety that produces beautiful, tight, round, red roots.

RHUBARB

FACTS/TIPS
- Rhubarb is a vegetable that we treat as a fruit.
- It is best raised by root division.
- It seldom suffers seriously from pests and diseases.
- The leaves are poisonous; cut them off and put them on the compost heap.

CALENDAR
Plant bare-rooted plants from late winter around to early spring; any time for container-grown plants.
Harvest – mid-spring to mid-summer.

SOIL AND SITE
- Rhubarb prefers a deep, rich, moist, well-drained loam in a warm, sheltered site.
- Soil that is either boggy and waterlogged or very dry is useless.
- In preparation, the soil should be double dug in the autumn, and enriched with plenty of manure.

SOWING AND PLANTING
- Plant divided roots in late winter to early spring in ground that has been previously double dug with plenty of rank manure, in a 1 ft deep hole that is wide enough to take the spread of the roots.
- Set the plants 30 in. apart.
- Fill around the root with a mix of topsoil and old manure.

PLANT CARE
Hoe the soil to prevent caking, and cover the ground with a mulch of

well-rotted manure. Water as often as possible. Remove flower stems as soon as they appear. Force the crop by covering the plants with straw and black plastic sheet.

HARVESTING
Harvest from late winter around to mid-summer, depending upon growing methods. Hold the stalk firmly and give it a half-turn tug so that it shears away from the crown. Trim the leaves off with a knife and put them on the compost heap.

VARIETIES
Glaskins Perpetual: Reliable, long-cropping variety that produces an abundance of bright red stems over a long period.
Hawkes Champagne: Old, tried-and-trusted, early variety that produces thick, rosy red stalks.
Victoria: Reliable, late variety.

RUTABAGAS

FACTS/TIPS
- Don't be tempted to grow giant varieties (unless you keep goats); it is much better to go for the smaller types.
- Rutabagas is so hardy that you can leave them in the ground over winter and lift and use them to suit.

CALENDAR
Sow – mid-spring to early summer.
Harvest – early autumn around to early spring.

SOIL AND SITE
- Rutabagas is happy in just about any soil, as long as it is deeply dug, moisture-retentive and fed with plenty of well-rotted manure.
- Avoid ground that looks to be sticky and/or puddled with standing water.
- Rutabagas does best in an open, semi-shaded, sunny position with shelter on the windward side.

SOWING AND PLANTING
- Sow seeds – mid-spring to early summer in ½ in. deep drills 10–12 in. apart. Water generously.
- When the seedlings are big enough

to handle, pinch out to leave the strongest plant 6 in. apart.
- Use your finger to firm the soil up around the remaining plants.

PLANT CARE
Hoe to create a loose-soil mulch. Be very careful that you don't graze the emerging root, as such damage might well result in top-rot or canker. Water little and often to avoid the wet-drought-wet conditions that bring on root splitting.

HARVESTING
Harvest from early autumn to early spring. Rutabagas is best when it is eaten young and tender. Store surplus roots by twisting off the leaves and burying them in peat in plastic dustbins.

VARIETIES
Angela: New early-cropping variety that produces medium to large, purple-skinned, creamy-fleshed roots.
Best of All: Hardy and productive variety that produces large, purple-topped, yellow-fleshed roots.
Marian: Hardy, vigorous and productive, disease-resistant variety that produces large, purple-skinned, yellow-fleshed roots.

SPINACH

FACTS/TIPS
- Freshly picked spinach is succulently tasty – not at all like the shop-bought variety.
- If spinach is good enough for Popeye, then fine!
- It is easy to grow.
- You can eat it raw in a salad.

CALENDAR
Sowing – early spring to early summer.
Harvest – late spring right around the year to the following spring.

SOIL AND SITE
- Spinach can be easily grown on almost any soil.
- It can be grown swiftly as a catch crop.
- While spinach will grow just about anywhere, it will only produce big,

plump leaves if the soil is well-prepared, well-manured and moisture-retentive.

SOWING AND PLANTING

- Sow seeds in early spring to early summer in 1 in. deep drills 10–12 in. apart. Sow thinly and cover.
- Compact the soil and use a fine spray to water generously.
- When the seedlings are big enough to handle, first pinch out to leave the strongest plants 3 in. apart, and then later thin it out to 6 in. apart.
- Water before and after thinning, and eat the thinnings.

PLANT CARE

Hoe to create a loose-soil mulch. Water liberally. If the weather becomes dry, spread a generous layer of spent manure mulch along both sides of the row and keep watering.

HARVESTING

Harvest from late spring to the following spring. Pick the leaves by hand, also keeping the plants in good condition by breaking off old and tired leaves.

VARIETIES

Atlanta: Very reliable summer- and autumn-cropping variety that produces thick, dark green leaves.
Giant Thick-leaved Prickly: Winter- and spring-cropping variety that produces huge, dark green leaves.
Giant Winter: Very popular and reliable, hardy, winter variety that produces a mass of large, dark green leaves.

SWEETCORN

FACTS/TIPS

- New varieties can, with lots of care, be grown in a relatively cold conditions.
- Sweetcorn is a good starter crop for interested children.
- Newer hybrid varieties must be grown away from each other to prevent cross-pollination.

CALENDAR

Sow – mid-spring to early summer.
Harvest – mid-summer to mid-autumn.

SOIL AND SITE

- Sweetcorn prefers a deeply worked, well-drained, light to sandy soil in a sheltered, sunny position.
- Dig plenty of well-rotted manure and compost in during the autumn.
- If you see the ground drying out, soak it with water and cover it with a mulch of old, well-rotted spent manure.

SOWING AND PLANTING

- Sow seeds in mid to late spring, with one chitted seed per peat pot under glass/plastic, and water generously.
- Plant in late spring to early summer. Prior to planting, harden the plants off by standing them outside in a sheltered position in the sun.
- Dig holes to fit your peat pots 12–18 in. apart in a grid pattern, and water the plants into place.

PLANT CARE

Hoe to create a loose-soil mulch. In dry weather, water and cover the ground with a thick mulch. Protect with a net when the cobs start to show.

HARVESTING

Harvest from mid-summer to mid-autumn when the tail-like silks are black-brown in color. Pick the cobs by hand, with a swift twist-and-down action.

VARIETIES

F1 Sweet Nugget: Reliable, outdoor, early autumn-cropping variety that produces medium-long cobs with large, sweet-tasting, golden kernels.
F1 Tasty Gold: Super-sweet early autumn-cropping variety that produces 9–10 in. long cobs with large, golden kernels.
F1 Tuxedo: Tall, disease-tolerant, sugar-enhanced variety that produces 9–10 in. long cobs with well-fattened kernels.

TOMATOES, INDOOR

FACTS/TIPS

- When my grandfather was a child, tomatoes were known as 'love-apples.'
- Plastic cloches or a polytunnel are a good option.
- Indoor tomatoes are more liable to suffer from mildew and mould than those grown outside.

CALENDAR

Sow – late winter to early spring.
Plant – early to late spring.
Harvest – early summer to mid-autumn.

SOIL AND SITE

- The easiest way to grow indoor tomatoes is in a growing-bag. You simply sit the bag in a sheltered, sunny position and then get on with the sowing and planting.
- Water as much as possible as long as the water does not puddle around the plants.
- Tomatoes hate irregular watering and stagnant water.

SOWING AND PLANTING

- Sow in late winter to early spring in a tray on a bed of moistened potting compost; protect with a sheet of glass and newspaper.
- Set out in mid to late spring. When the seedlings are large enough, prick them out into 3 in. peat pots.
- Plant in late spring to early summer in growing-bags, water generously, and place under cover.

PLANT CARE

Support the plant with a cane and loose ties. Pinch out the sideshoots that grow in the crutch between leaf and stem. Remove the growing point just short of the top of the shelter. Water with a dedicated liquid feed in the water. Remove yellow foliage.

HARVESTING

Harvest from early summer to mid-autumn. When the first bunch of fruits starts to ripen, strip off all the leaves from the ground to the first leaf above the bunch. Pluck the tomatoes when firm and nicely colored.

VARIETIES

Beefsteak: Good, tall-growing, indoor variety that produces large, pinky-red fruits.

Harbinger: Early-cropping variety that produces a heavy crop of sweet-tasting tomatoes.

Sun Baby: Intermediate variety that produces small, yellow, cherry-like fruits.

TOMATOES, OUTDOOR

FACTS/TIPS

- Outdoor tomatoes are generally healthier and more flavorsome than those grown indoors.
- Outdoor tomatoes are best planted in full sun with protection on the windward sides.
- Tomatoes hate irregular watering and stagnant water.

CALENDAR

Sow – early to mid-spring.
Plant – late spring.
Harvest – mid-summer to mid-autumn.

SOIL AND SITE

- Outdoor tomatoes can stand a light or heavy soil as long as it is deeply worked, richly manured, compact, well-drained, and in a sunny, sheltered position.

SOWING AND PLANTING

- Sow in early to mid-spring in a tray on a bed of moistened potting compost, and protect with a sheet of glass topped with newspaper.
- When the seedlings are large enough, prick them out into 3 in. peat pots. Water and keep warm.
- Plant in late spring to early summer. Set the peat pots in place in a sheltered position, and water generously.

PLANT CARE

Support the plant with a stake and loose ties. Pinch out the sideshoots. Remove the growing tip when there are 5–6 trusses. When the fruit starts to ripen, cover the ground with a mulch of rotted manure topped with a bed of straw, remove the supports, and gently lay the plant down.

Cover with the cloche and water the roots.

HARVESTING

Harvest from early summer to mid-autumn. When the first bunch of fruits is beginning to ripen, strip off all the leaves to the first leaf above the bunch, and remove damaged fruit. Check that the straw is crisp and dry. Pluck the tomatoes when they are firm and nicely colored.

VARIETIES

Gartenperle: Early to crop, bush-type variety that produces masses of small, cherry-like fruits.

Marmande: Very popular, late-maturing variety that produces large, uneven fruits.

Moneymaker: One of the most popular and reliable, heavy-cropping varieties of all time that produces masses of small to medium-sized fruits.

TURNIPS

FACTS/TIPS

- Some turnips are ready to lift in six weeks.
- Turnip tops can be eaten as spring greens.
- Varieties can be very different one from another; you need to choose carefully.

CALENDAR

Sow – late winter to early autumn.
Harvest – from mid-autumn right around the year.

SOIL AND SITE

- Turnips do well on a rich, light, sandy, moisture-retentive loam.
- If you have a choice between a sticky clay or dry sandy soil, opt for the sand and make adjustments.
- The soil needs to be deeply worked, well-manured and moist.
- Turnips do best in an open, semi-shaded, sunny position with shelter on the windward side.

SOWING AND PLANTING

- Sow seeds from late winter to early autumn in ½ in. deep drills 10–12 in. apart. Sow thinly and cover.

- Compact the soil and water generously.
- When the seedlings are big enough to handle, first pinch out to leave the strongest plants 3 in. apart and then later to 6 in. apart. Water before and after thinning.

PLANT CARE

Hoe to create a loose-soil mulch. Be careful not to damage the turnips' shoulders, as this might result in top-rot or canker. Water little and often.

HARVESTING

Harvest the roots from mid-autumn to early winter, and the tops in spring. Lift the roots as needed. Eat them when they are young and tender. To store surplus roots, twist off the leaves and bury the roots in peat in plastic dustbins.

VARIETIES

Golden Ball: Popular spring-sowing variety that produces sweet-tasting, golden, firm-fleshed roots.

Green Top Stone: Popular and reliable autumn-sowing variety that produces smallish, green-white roots.

Milan Purple Top: Very attractive, fast-maturing early variety that produces a flat-shaped, purple-topped, white-fleshed root.

A–Z OF FRUIT

APPLES

FACTS/TIPS
- Dessert apples are held in the hand and munched, and cooking apples are cooked.
- There are hundreds of varieties to choose from.
- There are free-standing trees, and trees that you train up against supports.

CALENDAR
Plant bare-rooted – autumn to spring.
Plant container-grown – all year.
Prune – late autumn to early spring.
Harvest – mid-summer to late autumn.

SOIL AND SITE
- Apples prefer a well-drained, moist, deep, loamy or brick-earthy soil.
- Double dig the ground to break through the hard subsoil.
- Dig in well-rotted farmyard manure and compost.

PLANTING
- Autumn to spring for bare-rooted trees, and any time for container-grown.
- Dig a 2 ft deep, 3–4 ft diameter hole.
- Set a bare-rooted tree in the hole, so that the topmost roots are 3–4 in. below the surface, and support with a vertical stake.
- Set a container-grown tree in the hole and support with an angled stake that points into the wind.
- Put a small amount of spent manure or compost into the hole and fill with topsoil.

PLANT CARE
Prune as the buds appear in spring or in autumn if you planted in spring. Cut the leading shoots back to half the length of the branch. With cordons, shorten sideshoots to within two or three buds of the stem.

HARVESTING
Pick the apples when they part readily from the spur. Remove with a slight pull-twist movement.

VARIETIES
Bramley's Seedling: The perfect cooking apple – green-skinned and white-fleshed.
Cox's Orange Pippin: Popular traditional variety that is firm-textured and sweet-tasting.
Discovery: Popular, early, disease-resistant variety that produces a bright red eating apple.

APRICOTS

FACTS/TIPS
- Apricots need just the right conditions; if you live in a cold climate, you either have to grow them indoors or grow them against a protected sunny wall.
- They are self-fertile – they don't need a partner.
- The tree is able to bear fruit on one-year-old wood as well as on older wood.

CALENDAR
Plant bare-rooted – late autumn.
Container-grown – early autumn.
Prune – late winter for training, and early to mid-summer for an established tree.
Harvest – when the fruit come readily from the tree.

SOIL AND SITE
- The apricot prefers a friable loam over a stony subsoil.
- The soil must be well drained but slightly moist.
- A good option is to prepare a pit lined with brick and mortar rubble, so that the subsoil is well drained.
- Ideally, you need to set the trees near/against a sunny wall so that they are protected on the windward side.

PLANTING
- Late autumn for a bare-rooted two-year old – dig a 2 ft deep, 3 ft diameter hole, break up the subsoil, put in a layer of broken brick, and top it with soil.
- Set the bare-rooted tree in place with the topmost roots no more than 3–4 in. below the surface, and support it with a stake.
- Fill the hole with loamy topsoil and tread firm.

PLANT CARE
For trees and espaliers, refer to plums (page 136). At blossom time, spread netting or fleece over the tree to keep off the frost. Remove the protection during the daytime when the conditions are favorable. Thin out crossing branches and remove suckers. In early spring, mulch the ground with spent manure.

HARVESTING
Gather the apricots in mid to late summer while they are still firm. It is best to eat them freshly picked.

VARIETIES
Moorpark: Popular variety that produces large, orange-fleshed fruits.
New Large early: Traditional variety – produces large fruits in late summer.

BLACKBERRIES

FACTS/TIPS
- Gathering wild blackberries is a traditional activity in the UK and parts of the USA.
- Blackberries can be grown more or less anywhere.
- There are many cultivated straight and hybrid varieties.

CALENDAR
Plant bare-rooted – from late autumn to early winter, or early to mid-spring.
Plant container-grown – all year.
Prune – Prune established plants in late winter.

Harvest – mid-summer to mid-autumn.

SOIL AND SITE

- Blackberries like an ordinary, deeply dug, well-manured, moist soil.
- For light and sandy ground, spread a thick mulch of well-rotted farmyard manure to hold in moisture.
- Blackberries do best when they are planted in rows that run from north to south, so that the midday sun shines down the length of the rows.

PLANTING

- In late autumn to early winter for bare-rooted canes, and any time for container-grown ones, dig a 9 in. deep, 2 ft wide trench.
- Spread 4 in. of well-rotted manure over the bottom of the trench and set the canes at 6 ft intervals.
- Fill the trench up with well-rotted manure topped with compacted soil.
- Build a 6 ft high post-and-wire support fence, with horizontal wires at 1 ft intervals.

PLANT CARE

After planting cut down the canes to about 9 in. above the soil, each cut above a strong, healthy bud. During the summer, train the young canes along all but the top wire. In the second year, train the new canes up through to the top wire. In the autumn, cut down all fruiting canes. Repeat the procedure in all following years.

HARVESTING

Pick the berries when they start to turn black, when they should come cleanly away from plug and stalk.

VARIETIES

Himalaya Giant: Reliable, vigorous, late-cropping variety that produces heavy crops of large, black, high-acid fruits.
Merton Thornless: Low-growing, thornless, late-cropping variety that produces large, fat fruits.

CHERRIES

FACTS/TIPS

- Cherry trees are hardy.
- Growing traditional varieties can be a problem; the trees are huge, the fruits are difficult to protect from raiding birds, and so on.
- In view of the bird problem, it is best to plant dwarf bush or fan varieties against a wall.

CALENDAR

Plant bare-rooted – in late autumn.
Plant container-grown – any time.
Prune – early spring for training and early to mid-summer for an established tree.
Harvest – when the fruits are ripe in early to late summer, depending upon variety.

SOIL AND SITE

- Cherries prefer a well-drained, loamy soil, like apricots and plums.
- The ideal site is an open plot that slopes to the face the sun at midday, so there is plenty of sunshine, with good shelter to the windward side.

PLANTING

- In mid to late autumn for a bare-rooted two-year-old bush, dig a 2 ft deep, 3 ft diameter hole. Break up the subsoil and top it with a thin layer of brick and mortar rubble.
- Set the bare-rooted bush in the hole, so that the topmost roots are no more than 3–4 in. below the surface, and support it with a stake.
- Fill the hole with loamy topsoil and tread firm.

PLANT CARE

For trees and espaliers, refer to apples and pears (see pages 131 and 136). For a bush, wait until the spring and cut the central stem down to 90 cm (3 ft) high. In the following spring, cut back half of the new growth to just beyond an outward-facing bud. Cut out all secondary shoots at their bases. In subsequent summers, thin out crossing branches and remove suckers.

HARVESTING

Gather the cherries while they still feel firm to the touch. It is best to eat them while they are fresh.

VARIETIES

E arly Rivers: Old variety that bears large, black fruits in early summer.
Morello: Self-fertile variety that produces large, black fruits in late summer.

CRANBERRIES, HEATHLAND

FACTS/TIPS

- Cranberries, once known in England and some parts of America as 'fenberries,' make very good sauces and cooling drinks.
- Traditionally, cranberry juice was thought to be good for 'women's conditions.'

CALENDAR

Plant bare-rooted – in autumn or spring.
Plant container- grown – any time.
Prune – as with gooseberries and currants.
Harvest – when the fruit starts to change color and soften.

SOIL AND SITE

- The cranberry needs a soft, wet, boggy soil (like a sump hole in a bog garden) – a nice mix of loam, sand and moss peat.
- A good option is to line a pit with plastic sheet, and to fill it with the soil mix as described.

PLANTING

- In autumn or spring for container-grown plants, prepare wet 'sump holes' and set plants about 12–18 in. apart.

PLANT CARE

Water the ground to keep it boggy.

HARVESTING

Gather the cranberries when they turn red and start to soften. They are not so good fresh, and are much better eaten in jellies and tarts.

VARIETIES

There are no named varieties.

CURRANTS, BLACK

FACTS/TIPS

• Blackcurrants are more popular than either red or white currants.
• They are much easier to grow and care for than either red or white currants.
• Blackcurrants are usually made into jam, while red and white currants tend to be made into sauces and jellies.

CALENDAR

Plant bare-rooted – in late autumn or late winter to early spring.
Plant container-grown – all year.
Prune – prune established plants in late autumn to early spring.
Harvest – mid to late summer.

SOIL AND SITE

• Blackcurrants prefer a generously manured, well-drained, light, sandy, loamy soil.
• Hard subsoil needs to be broken up and worked with plenty of farmyard manure or leafmould.
• Plant in a sunny spot at about 5 ft apart, so that there is a good circulation of air, and protection on the windward side.

PLANTING

• Late autumn to late winter or late winter to early spring for bare-rooted bushes, and any time for container-grown.
• Dig a hole that is plenty wide and deep enough for the roots to spread out.
• Spread about 4 in. of well-rotted manure over the bottom of the hole and set the bare-rooted bush in place. Place container-grown plants so that the fill is level with the soil.
• Fill the hole up with well-rotted manure topped with compacted soil.

PLANT CARE

After planting, cut all stems down to about 1 in. above soil level. In the following season, cut out all shoots that have produced fruits.

HARVESTING

Pick the fruits immediately the color has turned, while the berries are still firm and shiny. Pick off whole clusters rather than individual berries. In a good year, you might need to harvest twice a week.

VARIETIES

Baldwain: Popular, compact, medium-sized, late variety that produces medium sweet to tart berries – one of the best varieties for making blackcurrant jam.
Boskoop Giant – Blackcurrant: Large, vigorous, early variety that produces big, long bunches of large, sweet berries.

CURRANTS, RED AND WHITE

FACTS/TIPS

• Red and white currants are great for jellies, jams and wine.
• They are good for beginners because they are just about the hardiest of all hardy fruits.
• Red and white currants need to be looked after in much the same way as gooseberries (see page 134).

CALENDAR

Plant bare-rooted – mid to late autumn, or late winter to early spring.
Plant container-grown – all year.
Prune established plants – late autumn to early spring.
Harvest – mid-summer to mid-autumn.

SOIL AND SITE

• Red and white currants prefer a deeply dug, well-manured, well-drained, medium to heavy loam, in a sunny, airy spot that is sheltered on the windward side.

PLANTING

• In mid- to late autumn or late winter to early spring for bare-rooted bushes, and any time for container-grown ones, dig a 15–18 in. deep hole. Spread about 4 in. of well-rotted manure over the bottom of the hole and set the bare-rooted bush in place.
• Fill the hole up with a mix of topsoil and well-rotted manure topped with compacted soil.

PLANT CARE

After planting, cut back each main branch by about half. In the following autumn, cut back by half all the shoots that have formed in the year. At the end of the following season, shorten those shoots produced during the season by half, and clear out any shoots at the center.

HARVESTING

Pick the fruits as soon as they change color. In a good year, you might need to pick once or twice a week.

VARIETIES

Laxton's No. 1: Popular red currant that produces large bunches of medium-sized fruits.
White Verdailles: Good-flavored white currant that produces long, heavy trusses of large fruits.

FIGS

FACTS/TIPS

• Figs have been cultivated for many centuries.
• The flowers are never on show as they are hidden away at the center of the fruits.
• Figs are thought to be the most richly flavored and wholesome of all fruits.
• They are self-fertile – they don't need a partner.

CALENDAR

Plant bare-rooted – late autumn around to early spring.
Container-grown – early autumn.
Prune – late winter for training and early to mid-summer for an established tree.
Harvest – when the fruit come readily from the tree.

SOIL AND SITE

• Figs prefer a medium garden soil with mortar and rubble added.
• Beds are best prepared by lining a pit with brick and mortar rubble so that the subsoil is well drained and the roots confined.
• Ideally, you need to set the trees near/against a sunny wall so that they are protected on the windward side.

PLANTING
- In late autumn to early spring, dig a 2 ft deep, 3 ft diameter hole and line it with old rubble and mortar. Set the bare-rooted tree in the hole, so that the topmost roots are no more than 3–4 in. below the surface, and support with a stake.
- Fill the hole with topsoil; tread firm.

PLANT CARE
At blossom time, spread netting or fleece over the tree to keep off the frost. Remove the protection during the daytime when the conditions are favorable. Water generously when the fruits are developing. Use netting to keep the birds off.

HARVESTING
Gather the figs in late summer to early autumn, when the fruit sags downwards on the stalk. Eat them freshly picked and/or dry them in a warm cupboard.

VARIETIES
Brown Turkey: Good variety for outdoors that produces a medium-sized purple fig.
White Marseilles: Reliable outdoor variety that bears green, pear-shaped fruits.

GOOSEBERRIES

FACTS/TIPS
- Gooseberries are a good choice for beginners in that even a neglected bush will give a surprising amount of fruit.
- The cooked fruits are universally popular in the form of gooseberry jam.

CALENDAR
Plant bare-rooted – mid to late autumn, or late winter to early spring.
Plant container-grown – all year.
Prune established plants – late autumn to early spring.
Harvest – early to late summer.

SOIL AND SITE
- Gooseberries prefer a deeply dug, well-manured, well-drained, medium to heavy loam.

- Some varieties are susceptible to moulds and mildews when they are grown on heavy land.
- They need a sunny, airy spot that is sheltered on the windward side.

PLANTING
- In mid to late autumn or late winter to early spring for bare-rooted bushes, and any time for container-grown, dig a hole that is plenty wide and deep enough for the roots to spread out.
- Spread about 4 in. of well-rotted manure over the bottom of the hole and set the bare-rooted bush in place. Set container-grown plants so that the fill is level with the soil.
- Fill the hole up with a mix of topsoil and well-rotted manure topped with well-compacted soil.

PLANT CARE
After planting, cut back each main branch by about half. In the following autumn, cut back by half all the shoots that have formed in the year. At the end of the following season, shorten those shoots produced during the season by half, and clear out any shoots at the center.

HARVESTING
Pick fruits for cooking as soon as they start to change color. In a good year, you might need to pick once or twice a week.

VARIETIES
Black Velvet: Award-winning, mildew-resistant variety that produces heavy yields of dark red, grape-sized fruits.
Leveller: Reliable, yellow-green, dual eating-cooking variety that produces large, down-covered fruit.

GRAPES

FACTS/TIPS
- Grapes are native to the temperate regions of the Mediterranean and southern Europe.
- They can be grown in a heated greenhouse, but in the context of self-sufficiency outdoor grapes are the only sensible option.

CALENDAR
Plant container-grown outdoors – mid to late autumn or late winter to early spring.
Prune – spring and summer, and late autumn to early winter.
Harvest – early autumn to mid-winter.

SOIL AND SITE
- Grapes prefer a moist, well-drained, deep, loamy or brick-earthy soil, in a warm, sunny position, with protection on the windward side.
- They do best against a wall that faces the sun at midday.
- Double dig the ground to break through the hard subsoil, and add rubble for drainage, and plenty of well-rotted manure.

PLANTING
- Plant container-grown vines outdoors in mid to late autumn or late winter to early spring.
- Dig a 2 ft deep, 4 ft diameter hole.
- Put in a 6 in. layer of rubble topped with turves, set grass-side down.
- Set the vine in place, cover the roots with earth, and support with a vertical stake.
- Put a small amount of spent manure or compost into the hole and fill with topsoil.

PLANT CARE
Cut the young vine right back after planting. In the following year, cut the sideshoots back to two buds, and shorten the young leading growth. When the growth begins, train two laterals out to each side of the main rod. Continue in the same way in the following years, training out the four laterals in spring and summer, allowing them to flower and fruit, and then cutting them back in late autumn or early winter. In spring, spread a mulch of well-rotted compost around the plant.

HARVESTING
Picking will depend upon growing conditions and variety.

VARIETIES
Brandt: Popular variety that produces a heavy crop of small, sweet-tasting, purple grapes.

Chasselas: Traditional variety – a good option for an outside wall – that produces a medium-sized, pink-amber grape.

MELONS

FACTS/TIPS
- Melons are very dramatic to grow, like marrows and squashes.
- They can be grown outdoors in a warm climate, or in greenhouses, polytunnels, cloches and cold frames elsewhere.

CALENDAR
Sow – mid to late spring.
Plant – late spring to early summer.
Harvest – mid-summer to mid-autumn.

SOIL AND SITE
- Melons prefer deeply worked, well-manured, moist, well-drained soil.
- Dig a hole/trench, and fill it with manure topped with a mound of soil.
- Set the plants in place.
- The mound catches the sun, and ensures that the soil is well drained.
- Once the plants are hardened off and growing, cover the mound with a mulch to hold in the moisture.

SOWING AND PLANTING
- Sow seeds in mid- to late spring in peat pots under glass/plastic.
- In late spring to early summer, dig holes 1 ft deep and wide, in lines 4–5 ft apart, fill with well-rotted manure and mound over with soil. Set the peat pot plants into the top of the mound and water generously.

PLANT CARE
Pinch out the growing point after the fifth or sixth leaf. Once the sideshoots have formed, pinch out to select the four best shoots. Train the shoots out and pinch them off when you have about four melons per plant. Hoe to create a loose-soil mulch. Spread a mulch of spent manure around the plant. Keep watering and mulching throughout the season with grass clippings and spent manure.

HARVESTING
Harvest from mid-summer to mid-autumn. Eat them within a few days of picking.

VARIETIES
Sweetheart: Good choice for a cold frame; it produces green/gray-skinned, orange-fleshed fruits.

MULBERRIES

FACTS/TIPS
- Mulberries, not to be confused with 'Mull berries,' an old name for blackberries, grow on trees.
- At first glance, a mulberry looks like a large raspberry.
- Mulberries are great for wines, jams and various desserts.

CALENDAR
Plant bare-rooted – in late autumn to early winter, or best in early spring.
Plant container-grown – any time.
Prune – early autumn for training.
Harvest – when the fruits are ripe, spread sheets over the ground and shake the tree.

SOIL AND SITE
- Mulberries prefer a deeply worked, moist, well-drained, loamy soil that has been generously mulched with manure.
- The ideal site is in an enclosed garden with the tree set against a wall so that it faces the sun at midday.

PLANTING
- Plant bare-rooted trees in late autumn to early winter, or best in early spring.
- Plant container-grown trees at any time.
- Dig a 2 ft deep, 3 ft diameter hole and half fill it with the appropriate soil mix.
- Set the 3–5-year-old tree in place, so that the topmost roots are no more than 3–4 in. below the surface, and support it with a stake.
- Fill the hole with loamy topsoil and tread firm.

PLANT CARE
Prune as little as possible – no more than to shape. In subsequent summers, thin out crossing branches and remove suckers.

HARVESTING
Gather the berries when they start to fall from the tree. Eat them fresh or, best of all, make them into wine and jam.

VARIETIES
Black Mulberry: A classic, with wonderful taste, color and texture.

PEARS

FACTS/TIPS
- Planting an espalier pear against a wire support is a good option for a small garden.
- A happy, newly planted, 3–4-year-old, container-grown pear will bear fruit about two years after planting.

CALENDAR
Plant bare-rooted – mid to late autumn.
Prune – late autumn to early winter.
Harvest – mid-summer to late autumn.

SOIL AND SITE
- Although the preference is for a well-drained loam, a pear grown on a well-drained heavy soil will generally do better than one grown in either dry, light sand or dry, heavy clay.
- Pears like a warm, sheltered, frost-free corner, with protection on the windward side.

PLANTING
- In mid to late autumn for a bare-rooted, maiden, container-grown tree, dig an 18 in. deep, 3 ft diameter hole, break up the subsoil and top with a thin layer of broken brick.
- Set the tree in the hole, so that the topmost roots are 3–4 in. below the surface.
- Fill the hole with loamy topsoil and tread firm.

135

PLANT CARE

For a tree or bush, refer to the section on apples (see page 131). For an espalier, after planting build a small fence-like frame with wires. Prune the bare-rooted maiden down to about 15 in. During the summer, tie the shoots to the support wire. In the early winter, lower the arms to the horizontal position and tie in place. Repeat this procedure in following years.

HARVESTING

Pick fruits for cooking as soon as they start to change color and while they are still hard. Take the pear between cupped palms and gently lever it off.

VARIETIES

Conference: Popular, self-fertile, dessert variety that produces long, compact fruits.
Louise Bonne of Jersey: Very reliable, tried-and-trusted, old, partially self-fertile, dessert variety that produces medium-sized, red-flushed, greenish-yellow fruits.

PEACHES

FACTS/TIPS

- The peach belongs to the rose family (Rosaceae).
- Peaches are relatively easy to grow.
- In the context of self-sufficiency, it is a good idea to grow peaches against a sunny wall.

CALENDAR

Plant bare-rooted – early to mid-autumn.
Plant container-grown – all year round.
Prune – late autumn for training, and mid to late winter for thinning and true pruning.
Harvest – early to late summer.

SOIL AND SITE

- Peaches prefer a deeply dug, well-manured, well-drained, medium to heavy loam.
- The ideal site is against a tall, sunny brick wall, with protection on the windward side.

PLANTING

- In early to mid-autumn, dig a hole about 3 ft deep and 4 ft wide.
- Spread about 6 in. of brick rubble over the bottom of the hole and top it with thick turves, set grass-side down.
- Spread a mix of loam and mortar rubble over the turves and top it with charred garden refuse – wood and plant ash. Set a partially trained, three-year-old, bare-rooted, fan tree in place, and fill the hole with soil so that the topmost roots are covered by 2–3 in. of soil.

PLANT CARE

Protect the tree with netting or fleece in spring to protect from frost. Move the protection in the daytime. Use the protection again in summer to keep the birds off. In dry weather, water generously and spread a thick mulch of spent manure over the ground to keep in the manure. At the end of the fruiting season, cut the growths back to new shoots.

HARVESTING

Pick fruits for cooking as soon as they start to change color. Eat freshly picked and/or bottle or dry.

VARIETIES

Hales Early: Hardy, fertile variety that produces large crimson fruit.
Peregrine: Reliable cropper that produces a large fruit with a blushed crimson skin.

PLUMS

FACTS/TIPS

- Plums are the hardiest of the 'stone' fruits.
- You can be eating fresh plums from mid-summer to late autumn.

CALENDAR

Plant bare-rooted – from mid to late autumn.
Plant container-grown – any time.
Prune – early spring for training and early to mid-summer for an established tree.
Harvest – mid-summer to late autumn.

SOIL AND SITE

- Plums prefer a well-drained, light, loamy soil.
- Wet soil will produce lots of foliage and little fruit.
- The ideal site is an open plot that slopes towards the midday sun, with good shelter on the windward side.

PLANTING

- In mid- to late autumn for a bare-rooted two-year old bush, dig a 2 ft deep, 3 ft diameter hole. Break up the subsoil and top it with a thin layer of broken brick.
- Set the bare-rooted tree in the hole, so that the topmost roots are about 3–4 in. below the surface; support it with a stake.
- Fill the hole with loamy topsoil and tread firm.
- For trees and espaliers, refer to apples and pears (see pages 131 and 136).

PLANT CARE

For a bush, wait until the spring and cut the central stem down to 3 ft high. In the following spring, cut back half of the new growth to just beyond an outward-facing bud. Cut out all secondary shoots. In subsequent summers, remove crossing branches and suckers.

HARVESTING

Gather the plums as soon as the color changes, while they still feel firm to the touch. Pick the plums separately to avoid bruising.

VARIETIES

Purple Pershore: Good, disease-resistant, heavy-cropping variety that produces medium-sized, blue-purple, purple-fleshed fruits.
Victoria: The most popular and reliable dual eating-cooking, self-fertile variety; it produces large crops of big, fat, firm, juicy, red-gold fruits.

QUINCES

FACTS/TIPS

- It is best in the context of self-sufficiency to choose one of the pear-shaped varieties.
- The raw fruits taste terrible, but are delicious in marmalade and jellies.

CALENDAR

Plant bare-rooted – early autumn.
Plant container-grown – any time.
Prune – summer and autumn.
Harvest – mid to late autumn.

SOIL AND SITE

- The quince likes a moist soil in a sunny, open position that is protected on the windward side.
- Traditionally, they were grown near ponds in ground that is moist but not waterlogged.

PLANTING

- In early autumn, dig a 2 ft deep hole about 3–4 ft wide and half fill it with a mix of well-rotted manure, ordinary garden soil, leafmould and compost.
- Set the young tree in place, cover the roots with the soil mix, and tread firm.

PLANT CARE

Spread a mulch of manure around the tree to keep the soil moist and to deter weeds. After fruiting, as soon as the leaves fall in autumn, shorten shoots to 3–4 buds of the base, and thin out crossing shoots.

HARVESTING

Gather fruits in early to late autumn when they change color and start to give off a characteristic scent, when they more or less drop into the hand when touched. If there is a danger of an early frost, pick them and store them in a cool dark cupboard.

VARIETIES

Champion: Popular variety that produces large, green-gold, apple-shaped fruits.
Pear Quince: Produces a prolific crop of large, pear-shaped fruits.

RASPBERRIES

FACTS/TIPS

- The raspberry is a hardy, deciduous shrub, with summer- and autumn-fruiting varieties.
- The harvesting window is small.

CALENDAR

Plant bare-rooted canes – late autumn to early winter.
Plant container-grown – any time.
Prune – immediately after planting and after fruiting.
Harvest – mid-summer to mid-autumn.

SOIL AND SITE

- Summer-fruiting raspberries need a well-drained, light soil, in a sunny, sheltered position.
- Give a good mulch of rotted manure in times of drought.
- Raspberries can be planted in a heavy clay soil, as long as the ground is free from standing water.
- Autumn-fruiting raspberries need more sun and extra shelter.

PLANTING

- In late autumn to early winter for bare-rooted canes, double dig the ground with plenty of well-rotted manure.
- Dig a trench 6 in. deep and 18 in. wide.
- Set the bare-rooted canes 15–18 in. apart in rows 5–6 ft apart.
- Fill the trench with a mix of topsoil and old manure and tread firm.

PLANT CARE

Build a support with horizontal wires 2 ft, 3 ft and 5 ft from the ground. After planting, cut all the canes down to about 1 ft. When new canes start to grow, cut the old 1 ft canes down to soil level. Tie the new canes to the support wires. After fruiting, cut all fruiting canes down to soil level, and tie up all the new canes.

HARVESTING

Harvest on a sunny day, as soon as the fruits are evenly colored. Snip them off complete with the stalk. Burn any damaged maggoty berries.

VARIETIES

Malling Orion: Early, heavy-cropping, summer-fruiting variety that produces medium-sized, pinky fruits.
Zeva: Hardy, autumn-fruiting variety that produces large, red fruits.

STRAWBERRIES

FACTS/TIPS

- Strawberries do best when they are grown en masse.
- They can easily be propagated by layering the runners into pots or straight into the ground.

CALENDAR

Plant bare-rooted plants – mid-summer to early autumn.
Plant container-grown – any time.
Prune – Straight after harvesting.
Harvest – late spring to mid-summer.

SOIL AND SITE

- Strawberries prefer a deep, heavy loam that is inclined to clay.
- The ideal is a site that slopes down to the sun at midday, with protection on the windward side.
- The soil needs to be deeply dug with lots of farmyard manure.

PLANTING

- Double dig the ground with plenty of well-rotted manure.
- In mid-summer to early autumn for bare-rooted plants, dig shallow holes wide enough to take the roots at full spread, in rows 2 ft apart, with the plants 18 in. apart in the rows.
- 'Puddle' the plants in place, top up with friable soil, and firm the soil.

PLANT CARE

When the strawberries are beginning to form, spread a thick mulch of clean, closely packed straw around the plants to keep them warm and clean. Push in sticks or wires to make a low bow-shaped support, and spread a net to keep the birds off.

HARVESTING

Pick the berries as they ripen, in the early morning when dry. Pick them complete with the stalks and plugs.

VARIETIES

Cambridge Favorite: Heavy-cropping, virus- and disease-resistant, early variety that produces medium-sized, orange-red fruits.
Honeoye: Heavy-cropping, early summer-fruiting variety that produces dark red, slightly soft fruits.

137

A–Z OF HERBS

BASIL

FACTS/TIPS
- If you enjoy Italian food, you need to grow basil.
- Basil is tender and is best grown under cover or on a windowsill.
- It is a half-hardy annual.

CALENDAR
Sow – under glass in early to late spring.
Harvest – mid to late summer.

SOIL AND SITE
- Basil does best in a rich, well-drained soil in a sunny, sheltered position.
- A sunny border not too far away from the kitchen is ideal.

SOWING AND PLANTING
- Sow seeds in early spring to mid-summer in boxes filled with compost and loam in equal parts. Sow thinly and cover with glass.
- When the seedlings are big enough to handle, first pinch out to leave the strongest plants, and then plant out in a windowbox, pots or a border, in a similar soil.

PLANT CARE
Hoe to create a loose-soil mulch. Water liberally. Pinch out the tips to produce a bushy plant. If the weather becomes dry, then spread a generous layer of spent manure mulch over the soil and keep watering.

HARVESTING
Harvest from mid to late summer. The leaves can be crushed, dried and stored in airtight bottles, or crushed and preserved in ice cubes.

VARIETIES
Bush Basil: Low-growing type that can be kept indoors; very good if you are short of space.
Sweet Basil: The most popular type; it produces a mass of bushy leaves.

BAY

FACTS/TIPS
- Bay is a hardy evergreen shrub with dark green, aromatic leaves.
- In ideal conditions, it can grow to a height of 6–12 ft.
- The leaves give a rich flavor to stews and fish dishes.
- The Greeks and Romans twined bay leaves together to make their wreaths or victory crowns.

CALENDAR
Propagate – plant cuttings in mid-to late summer.
Harvest – young leaves in spring and summer as needed.

SOIL AND SITE
- Bay does best in a deeply dug, well-drained, loamy soil in a warm, sheltered position.
- Traditionally, bay bushes or trees were grown in tubs and clipped into 'lollipop' trees. They are often clipped into a pyramid or cone.
- Bays are often kept outdoors in the summer and brought into a porch or conservatory in the winter.

SOWING AND PLANTING
- In early to mid-winter, set cuttings of young shoots in peat pots containing a sandy soil.
- Water generously and keep warm.
- In a few weeks, the cuttings will have made roots. Transplant the best examples into larger pots.
- Put them in a warm position with protection on the windward side.

PLANT CARE
Water a little and keep warm. Clip a couple of times a year, and wash with a spray hose to keep the leaves clean. Bring indoors if the weather turns cold.

HARVESTING
Harvest the young leaves as and when needed. Use leaves freshly picked and/or dry the leaves on the windowsill and store them in airtight containers for winter use.

VARIETIES
Sweet Bay: The sweet bay is an evergreen shrub, a bit like laurel. Cuttings and pot-grown examples are sold under the name Sweet Bay and/or *Laurus nobilis*.

CHERVIL

FACTS/TIPS
- Chervil is a hardy biennial that is usually grown as an annual.
- With bright green fern-like leaves, it looks a bit like parsley, and tastes like a cross between parsley and fennel.
- It grows to a height of about 18 in.
- The leaves have a delicate aniseed flavor – very good in salads and sandwiches, and with fish or eggs.

CALENDAR
Sow – late summer through to late winter and various intervals throughout the year.
Harvest – mid-summer to mid-autumn.

SOIL AND SITE
- Chervil prefers a well-drained, moderately rich soil.
- Plant summer crops in a cool, sheltered border and winter crops in a sunny border.

SOWING AND PLANTING
- Sow seeds in late summer in drills 8–10 in. apart in ordinary garden soil in a sheltered, cool border.
- Sow seeds in winter in well-drained, light soil, in seed boxes under glass.
- Thin the seedlings out to about 5 in. apart.
- Plant them out, as soon as they are large enough to handle, in large pots or directly into a border (cool border for summer planting and warm border for winter planting).
- The leaves are more tender and have a better flavor when the plants are grown swiftly in succession. For maximum flavor, grow new plants every year.

PLANT CARE

Hoe to create a loose-soil mulch. In dry weather, water and cover the ground with a thick mulch. Protect young plants with a net or fleece.

HARVESTING

Harvest year round as and when needed. Young leaves can be plucked and used directly. Let some plants run their course so that you have seeds for next year.

VARIETIES

Chervil Plain: The leaves are longer, straighter and stronger-tasting than the curly variety.
Curly Chervil: The fern-like leaves give a delicate, aniseed-like taste that is very good for mild-flavored dishes.

CHIVES

FACTS/TIPS

- Chives are hardy, low-growing, clump-forming perennials, with green, tubular stems topped with round, rose-pink flowerheads.
- The chopped-up stems or leaves have a beautifully distinctive, tart, onion-like flavor – good for sandwiches and omelettes.
- Chives look really good as an edging to a lawn or to a raised border.

CALENDAR

Sow – outdoors early to mid-spring.
Plant – early to late spring.
Harvest – cut young and tender leaves when needed. Cut away the whole clump, whether the leaves are needed or not, so as to achieve a regular supply of fresh, young leaves.

SOIL AND SITE

- Chives prefer a rich, moist soil in a warm, sunny, sheltered position.
- Water generously and regularly in dry weather.

SOWING AND PLANTING

- Sow in early spring in ½ in. deep drills in rows 1 ft apart.
- Set out in mid- to late spring; thin out to about 3 in. apart.

PLANT CARE

Hoe to keep down the weeds and to achieve a loose-soil mulch. Water generously. In very dry weather, cover the ground with a mulch of spent manure. Divide the clumps every 3–4 years and/or plant afresh. Leave some plants for their attractive flowers, but for those plants intended for culinary use pick and remove the flowerheads before they have a chance to open.

HARVESTING

Cut the grassy leaves as needed.

VARIETIES

Border Chives: The thin, grass-like leaves have a mild onion flavor that is very good for salads; also looks good as an edging to a border.
Windowsill Chives: Good as a windowbox plant and/or for greenhouse forcing; it produces thick, dark, strong-flavored leaves.

DILL

FACTS/TIPS

- Dill is a hardy annual with tall stems topped with feathery, blue-green leaves that grow to a height of about 2–3 ft.
- The freshly picked leaves can be used to best effect to garnish and flavor new potatoes and white fish.
- You can harvest the leaves and the seeds; both have a distinctive flavor, a bit like mild aniseed.
- Oil from the seeds has carminative, stimulant and aromatic properties.

CALENDAR

Sow – early to mid-spring.
Harvest – early to mid-summer and then as needed.

SOIL AND SITE

- Dill does well on an ordinary, well-drained, moderately fertile soil in a sunny, sheltered position.
- Dill comes to a halt if it is disturbed, so sow the seeds in the final planting position.

SOWING AND PLANTING

- Sow seeds in early to mid-spring in shallow drills about 1 ft apart. Sow thinly and cover.

- Compact the soil and water generously.
- When the seedlings are big enough to handle, thin out to leave the strongest plants about 1 ft apart.
- Water before and after thinning.

PLANT CARE

Hoe to create a loose-soil mulch. Water little and often.

HARVESTING

Harvest in early to mid-summer and then as needed. When the fruits are ripe, cut the whole stems and spread them out in a warm, dry, sheltered corner of a shed or conservatory. When the heads are crisp, dry the heads and gather the seeds. The leaves can be gathered as needed, and/or dried and stored in lidded containers.

VARIETIES

Bouquet: Dwarf variety that produces deep blue-green leaves on a compact bush.
Dukat: Vigorous variety that produces blue-green leaves and is noted for its high oil content and drying qualities.
Hercules: Large plant that produces masses of highly flavored leaves – a very good option for drying.

FENNEL

FACTS/TIPS

- Fennel is a hardy herbaceous perennial with tall stems, feathery, green leaves, and golden-yellow flowerheads.
- It grows to a height of 5–6 ft.
- The leaves are good with fish, salads and stews; the seeds add taste and texture to cakes, bread and soups.

CALENDAR

Sow – early to mid-spring.
Harvest – pick the leaves in summer as needed.

SOIL AND SITE

- Fennel does best in a moist, well-drained, moderately fertile soil in a sheltered, sunny position.

- The tall plants are decorative, so could be sown in place in a sheltered border.
- Tall plants will need support.

SOWING AND PLANTING

- Sow seeds in early to mid-spring in shallow drills about 18 in. apart. Sow thinly and cover.
- Compact the soil and water generously.
- When the seedlings are big enough to handle, thin out to leave the strongest plants about 1 ft apart, and/or transplant.
- Water before and after thinning and transplanting.

PLANT CARE

Hoe to create a loose-soil mulch. Water little and often. Tall plants in a windy spot will need to be supported with canes. Control the height by picking out the topmost shoots. Remove the flowerheads to encourage leaf growth.

HARVESTING

Harvest in early to mid-summer and then as needed. The leaves can be gathered as needed, and/or dried and stored in lidded containers. The stems can be sliced thinly and used in salads.

VARIETIES

Fennel Sweet: Tall, feathery plant that produces tender stems and leaves.
Fennel Sweet Bronze: Much the same as Fennel Sweet, apart from the beautiful bronze color.

GARLIC

FACTS/TIPS

- We all know about keeping vampires and evil spirits at bay, but garlic was also chewed by athletes in ancient Greece and Rome, and worshipped in ancient Egypt.
- Garlic is a hardy bulbous perennial of the onion family.
- It is still considered to be one of the 'wonder' foods – good for just about anything from a cold through to weight loss.
- It is easy to grow and just as easy to store.

- Garlic is very good for flavoring meat, and better still in a cheese sandwich.

CALENDAR

Sow – mid-winter right around to mid-autumn, depending on variety, shelter and growing methods
Harvest – mid-summer to mid-autumn.

SOIL AND SITE

Garlic does best in a rich, light, well-drained soil in a sunny, sheltered position.

SOWING AND PLANTING

Plant cloves in early spring to mid-summer, or mid-autumn to late winter if you want an early crop, in 2–3 in. deep drills, 6–8 in. apart, in rows 1 ft apart.

PLANT CARE

Hoe to create a loose-soil mulch and to keep the weeds at bay. Water liberally. Keep sheltered for winter planting.

HARVESTING

Harvest from mid to late summer for main crop. When the foliage turns yellow, lift, leave to dry and hang in a very dry shed.

VARIETIES

Common White: Produces small to large heads; the outer skin is a silvery-white color.
Pink: Produces heads similar to the Common White, but crops earlier; the skin is a pinky color.
Red: Has a red skin and much larger and flatter cloves than the white and pink varieties.

HORSERADISH

FACTS/TIPS

- Horseradish is a hardy perennial with a single, long root, a bit like a dandelion or a slender parsnip.
- It is the perfect accompaniment to roast beef or salmon.
- Although horseradish is sometimes left undisturbed for years, the best practice is to lift and plant new every other year.

CALENDAR

Plant – plant root cuttings in early winter around to early spring.
Harvest – roots from summer to late autumn as needed, and store in a frost-free shed.

SOIL AND SITE

- Horseradish does well in just about any ordinary garden soil – sunny or shaded – as long as it is fertile and well-drained.
- Choose a corner that you know can remain undisturbed for a couple of years.
- The trick is to plant it so that it is contained, for example in a plastic water butt set into the ground.

SOWING AND PLANTING

- Plant root cuttings in early winter to early spring. Cut medium-thick roots into pieces about 6–10 in. long and set them into 3 in. deep holes about 9 in. apart. Water generously and keep away the weeds.

PLANT CARE

Hoe the soil to keep off the weeds and to create a loose-soil mulch. Water regularly.

HARVESTING

Harvest the roots from summer through to mid-autumn and store in a cool, dry, dark place.

VARIETIES

There are no named varieties.

MARJORAM, SWEET

FACTS/TIPS

- Sweet marjoram is a slightly tender perennial that is grown as a half-hardy annual.
- Sweet marjoram and oregano are the same, and are also known as winter marjoram, annual marjoram and pizza oregano.
- Seeds are sown in late winter or early spring, and the leaves are gathered as needed.
- The leaves are very good for flavoring stews and soups.
- It grows to a height of about 2 ft.

CALENDAR
Sow – late winter to mid-spring.
Harvest – as needed.

SOIL AND SITE
- Marjoram prefers a well-drained, sandy soil in a sunny, sheltered position.
- It is even better if the soil is enriched with a light-textured spent manure.

SOWING AND PLANTING
- Sow seeds under glass in early spring in trays filled with a light, sandy soil.
- Thin the seedlings out to about 3 in. apart, and plant them on either to deep trays or pots.
- Plant out in mid-spring in a sunny, well-drained position. Set them 9–12 in. apart in rows 1 ft apart.

PLANT CARE
Hoe to create a loose-soil mulch. In dry weather, water and cover the ground with a thick mulch. In very hot weather, shade the young plants with mesh.

HARVESTING
Harvest the leaves when young and as needed. At the end of summer towards early autumn, either strip off bunches of leaves or pull up the plants just before flowering, and hang in a cool, dry, shaded shed to dry. Finally, save the dried leaves for winter use.

VARIETIES
Pot Marjoram: Dwarf perennial, grown as an indoor or outdoor pot plant, that is easier to grow than sweet marjoram.

MINT

FACTS/TIPS
- Common mint is a hardy herbaceous perennial with mid-green leaves
- It grows to a height of about 2 ft.
- Mint will grow just about anywhere and can be extremely invasive.

- The leaves are perfect when chopped with brown sugar and used with vinegar in the form of a mint sauce.

CALENDAR
Plant – spring or autumn.
Harvest – as needed.

SOIL AND SITE
- Mint likes a moist soil in a warm or dappled shade position.
- Water generously in dry weather and topdress with compost in autumn.
- Mint is invasive and can become a nuisance; a good option is to sink a plastic water butt in the ground before planting, so that the roots are contained.

SOWING AND PLANTING
- Plant root clumps every other spring or autumn. Lift the old roots and use a spade to split them into pieces. Plant the outer edges and throw away the old central parts.

PLANT CARE
Hoe to keep down the weeds and to achieve a loose-soil mulch. Water generously in dry weather and topdress with compost in autumn. In very dry weather, cover the ground with a mulch of spent manure.

HARVESTING
Pick the young leaves as and when needed. To obtain a supply for winter months, you can dry and store the leaves, and/or you can lift a plant, set it in a deep box of loamy sand and keep it in a greenhouse.

VARIETIES
Emperor's: Originally found around the Emperor Hadrian's villa near Rome, this has a mint-like flavor and fragrance.
Peppermint: Has long spikes of mauve flowers in autumn.
Spearmint: Also called Green Pea Mint, this has a refreshing spearmint flavor that is good with roast lamb.

PARSLEY

FACTS/TIPS
- Parsley is a hardy biennial that tends to be grown as an annual.
- It has lots of curly, tight-packed, green leaves.
- The leaves are commonly used in sauces and as a garnish.

CALENDAR
Sow – early spring to mid-summer.
Harvest – as needed.

SOIL AND SITE
- Parsley prefers a crumbly, moisture-retentive, fertile soil containing lots of friable leaf mould, in full sun on heavy ground and in shade on light ground.
- If your parsley looks a bit tired, topdress with leafmould, and next time around add friable leafmould with the soil at the sowing stage.

SOWING AND PLANTING
- Sow seeds in early spring or mid-summer in ½ in. deep drills about 12–14 in. apart. Sow thinly and cover.
- Compact the soil and water.
- When the seedlings are big enough to handle, thin out to leave the strongest plants about 8–9 in. apart.
- Water before and after thinning.

PLANT CARE
Be aware that parsley is very slow to germinate. Hoe to create a loose-soil mulch. Water little and often. If the weather is hot, topdress with friable leafmould.

HARVESTING
Harvest fresh sprigs in summer and winter as needed. To obtain a supply for the winter months, you can dry and store the leaves, and/or you can lift a plant, set it in a deep box of friable soil and keep it in a greenhouse.

VARIETIES
Champion: Has a rich green, closely curled, compact habit.
Forest Green: Has longer, stiffer stalks and very dark green leaves – a very popular variety.

Hamburg: Variety grown for its roots rather than the leaves. Tastes a bit like a mix between celery, parsley and parsnips – very tasty in salads or mashed and fried.

ROSEMARY

FACTS/TIPS
- Rosemary is an evergreen shrub with narrow, spiky, mid to dark green, needle-like, aromatic leaves.
- It grows to a height of 6–7 ft.
- It has a very strong flavor and aroma and was traditionally thought to have a stimulating effect on the mind; it was also used as a disinfectant, and as such was burnt in sick rooms.
- The leaves are used to flavor all manner of fish and meat dishes.
- The tall plants are both decorative and fragrant, so were traditionally planted near a pathway so that passers-by could pick the leaves.

CALENDAR
Sow seeds – in late spring.
Take cuttings – late summer and autumn.
Harvest – pick as needed.

SOIL AND SITE
- Rosemary will grow just about anywhere but does best in a well-drained, light soil in a sunny position.
- Heavy soil can be made light and friable by mixing it with sand and leafmould.

SOWING AND PLANTING
- Sow seeds in late spring in shallow drills about 18 in. apart. Sow thinly and cover.
- Compact the soil and water generously.
- Plant cuttings in late summer and autumn. Place 9 in. long sections of sideshoot with the heel intact 5 in. deep in a cold frame. In the following year, set the rooted plants 1 ft apart in a bed. In year three, plant them out in their final positions.

PLANT CARE
Hoe to create a loose-soil mulch. Water little and often. Prune annually after flowering, and cut the roots back to create a compact, bushy shape.

HARVESTING
Harvest fresh as needed. In readiness for winter, the leaves can be gathered and dried and stored in lidded containers.

VARIETIES
The species, *Salvia officinalis*, is the most commonly grown.

SAGE

FACTS/TIPS
- Sage is a hardy, evergreen shrub with long, aromatic, green-gray leaves.
- It grows to a height of about 2 ft.
- The leaves are extensively used in a whole range of dishes – everything from sage and onion stuffing to cheese-sage dip.
- Sage has a very strong, bitter flavor – it is one of the strongest herbs.

CALENDAR
Cuttings – in early autumn.
Harvest – gather young leaves as needed.

SOIL AND SITE
- Sage prefers in a well-drained, light, fertile soil in a sunny, sheltered position.
- Keep the soil moist.

SOWING AND PLANTING
- Plant cuttings in early spring through to early autumn. Set 3–4 in. long 'slips' or top-shoot cuttings about 4 in. apart in a loam-based compost in a cold frame.

PLANT CARE
Hoe to create a loose-soil mulch and to keep down weeds. Water liberally. Trim back annually to prevent the plant becoming open and straggly.

HARVESTING
Harvest the young leaves as need. For drying, collect the foliage before flowering, dry in a warm room, and store in airtight jars.

VARIETIES
Blue Sage: An attractive, half-hardy, tuberous-rooted variety with green-gray leaves and deep blue flowers.
Common Sage: Has green-gray leaves and a bitter taste. Also called Common Garden Sage.

SAVORY, SUMMER

FACTS/TIPS
- Summer savory is a bushy annual, unlike winter savory, which is an evergreen perennial that can be grown as an annual; they can both be used as herbs.
- It has slender, erect, branching stems and small, narrow, pale green leaves
- It grows to about 1 ft in height.
- Savory is particularly good in soups, stews, and meat and fish dishes.
- Traditionally, savory leaves were rubbed on to ease the pain of bee stings.

CALENDAR
Sow – mid to late spring.
Harvest – the leaves are gathered fresh as needed, also pulled and dried when coming into flower.

SOIL AND SITE
- Summer savory prefers a deep, fertile soil in a sunny, sheltered position.

SOWING AND PLANTING
- Sow seeds – in mid to late spring – in 1/2 in. deep drills 8–12 in. apart.
- Water generously and keep away the weeds.
- When the seedlings are large enough to handle, thin them out to the best plants 6–8 in. apart.

PLANT CARE
Hoe the soil to keep off the weeds and to create a loose-soil mulch. Water regularly. In very dry weather, cover the ground with a mulch of spent manure.

HARVESTING

Harvest the leaves fresh, as needed. Pull the plants up when they are coming into flower, tie them into bundles and hang to dry in a cool, airy shed. Pick the dry leaves and store them in airtight containers.

VARIETIES

Summer Savory: All as described above.
Winter Savory: Perennial shrub with small, narrow, pointed leaves and pale lilac flowers.

SORREL

FACTS/TIPS

• Sorrel (also known as French Sorrel, Bread and Cheese, Sour Dock and many other names) is the plant that many people would know and recognize as the common Dock.
• It grows over 3 ft in height.
• Traditionally, the tall-growing variety was cooked and eaten in much the same way as spinach.
• Sorrel (Rumex acetosa) is eaten as a salad, while the similar species Rumex patienta (also known as Patience) is eaten like spinach.

CALENDAR

Sow – mid to late spring.
Harvest – from early spring and as needed.

SOIL AND SITE

• Sorrel prefers a well-manured, moist soil in a warm to dappled shade position.
• Water generously in dry weather and topdress with compost in autumn.
• Sorrel is of course Dock and can be invasive. A good option is to sink one or more plastic water butts in the ground before planting so that the roots are contained.

SOWING AND PLANTING

• Sow seeds from mid- to late spring in 1 in. deep drills in lines about 1 ft apart. Thin first to 6 in. and then to 18 in. apart.

PLANT CARE

Hoe to keep down the weeds and to achieve a loose-soil mulch. Water generously in dry weather and topdress with manure in autumn.

HARVESTING

Pick the young leaves as and when needed.

VARIETIES

Rumex acetosa (Sour Sorrel):
Good as a salad. A bit acid or sour in taste – hence the common name.
Rumex patienta (Patience):
Good as cooked greens, just like spinach.

TARRAGON

FACTS/TIPS

• Tarragon is an evergreen shrub grown for its narrow, spiky, mid to dark green, needle-like, aromatic leaves.
• The woody stems grow to about 3 ft in height.
• The leaves are used as a flavoring in tarragon vinegar and tartare sauce.
• It is good with all sorts of foods – the mint-like taste and aroma is good especially with white fish.

CALENDAR

Plant roots – in mid to late spring.
Harvest – pick fresh as needed and/or pick and dry in autumn.

SOIL AND SITE

• Tarragon does best in a well-drained, light, garden soil in a sunny position.
• In cold districts, provide shelter on the windward side.

SOWING AND PLANTING

• Plant pieces of root (rhizomes) in mid to late spring in 3–4 in. deep holes, 12–15 in. apart.
• Cover, compact the soil, and water generously.

PLANT CARE

Hoe to create a loose-soil mulch. Water little and often. As the flowers appear, pinch out the flower stems to boost leaf production.

HARVESTING

Harvest fresh in early to late summer as needed. In readiness for winter, the leaf buds are cut, dried and stored in airtight containers.

VARIETIES

French Tarragon: Also known as Estragon and simply as Tarragon. There are no other named varieties.

THYME

FACTS/TIPS

• Thyme is a hardy, dwarf, evergreen shrub with small, aromatic leaves.
• Thyme is the common name for a whole range of low-growing plants.
• The leaves are used with fish, and with rich meats such as hare and pork.

CALENDAR

Plant roots – in mid to late spring.
Harvest – pick fresh as needed and/or pick and dry in autumn.

SOIL AND SITE

• Thyme prefers a well-drained, light garden soil in a sunny position.

SOWING AND PLANTING

• Plant pieces of root in mid to late spring in 3–4 in. deep holes, 9–12 in. apart.
• Cover, compact the soil, and water generously.

PLANT CARE

Hoe to create a loose-soil mulch. Water little and often.

HARVESTING

Harvest fresh in early to late summer as needed. In readiness for winter, cut off long stems and dry them in a dry, airy shed.

VARIETIES

Common Thyme: The best one for use in the kitchen, and it also makes a refreshing tea.
French Thyme: Also known as Summer Thyme; it has a stronger and sweeter flavor than Common Thyme.

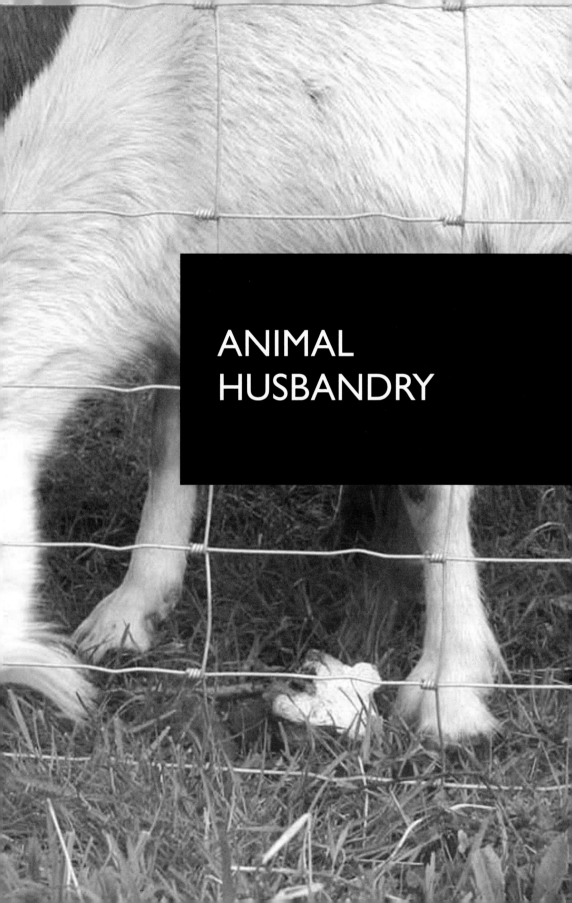

ANIMAL
HUSBANDRY

KEEPING CHICKENS

Chickens and self-sufficiency go together like boiled eggs and bread and butter, or bacon and eggs. Every morning you will be presented with fresh eggs – warm, tasty, beautiful, and just about as clean and green as food can get! Chickens are relatively easy to keep, and are therefore perfect for a small town garden. Chickens will recycle most of your kitchen scraps, help keep rough grass and weeds down, and eat lots of bugs and pests. You do have to feed them supplementary protein, and they do need looking after, but you will have fresh eggs for the best part of the year. Better yet, if you have say six hens and one cock, then before you know it you will also have chicks. If and when you want to kill a chicken, seek advice from a registered chicken society.

BREEDS

Each chicken breed has its own special qualities – a good reliable layer, lays brown eggs, small body size, hardy, large with firm white flesh, and so on. When choosing, look at your plot, decide on your needs and take it from there.

HOUSING

The chicken house must be dry, warm, easy to clean, and as large as space allows. There are lots of options, including everything from little plastic dome-shaped shelters – just right if you only want to keep a few birds in a small garden – through to wooden huts and arks in just about every shape and size that you can imagine. My preferred option is a good-sized, traditional, walk-in-type shed, meaning a shed with some sort of vents or windows at the front, and a bank of nesting boxes at the side or rear. The outside door to the nesting boxes

GOOD EGG-LAYING BREEDS

RHODE ISLAND RED Traditional, robust, hardy breed with dark brown to red feathers and yellow flesh – a good choice if you want large brown eggs.

ORPINGTON Big, strong, hardy, friendly bird with buff to blue-black feathers and white flesh; it produces brown eggs.

PLYMOUTH ROCK Small, robust breed with partridge-blue feathers, barred white, and yellow flesh; very popular in America; it produces large, pink-to-brown eggs.

LIGHT SUSSEX Strong, tolerant bird with white feathers and creamy flesh; good for meat and eggs.

ISA BROWN Good-sized laying hybrid with brown to gold plumage; very high egg yield; produces good-sized brown eggs; a good choice if you want hardy birds.

ANCONA BANTAM Miniature version of the full-sized Ancona; produces very attractive small white eggs; slightly temperamental character.

BARRED PLYMOUTH ROCK Large, robust, hardy bird with barred plumage (dark markings on a grayish background); produces good-sized brown eggs.

WHITE LEGHORN Medium-weight bird with attractive white plumage; great layer, good for both farms and gardens; produces a large number of good-sized white eggs.

EXCHEQUER LEGHORN Traditionally good choice for smallholdings and gardens; medium-sized bird with black and white plumage; lays a large number of white eggs.

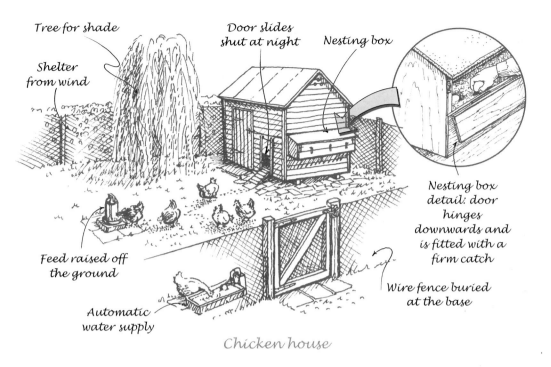

Tree for shade

Shelter from wind

Door slides shut at night

Nesting box

Nesting box detail: door hinges downwards and is fitted with a firm catch

Feed raised off the ground

Automatic water supply

Wire fence buried at the base

Chicken house

allows you to remove the eggs without going to all the mess and bother of entering the shed.

S ay that you want to keep six birds in a medium-sized back garden. Start by ringing your chosen area with close-mesh chicken wire. Position the chicken house within the enclosure so that the front is looking towards the midday sun. If you use mesh that is about 8 ft wide and bury about 18 in. of it in the ground, you will finish up with a fence that is 6–7 ft high. If you can include in this scenario a bush or low tree for shade, all the better. If you are worried about foxes either cover the whole run with a ceiling of wire mesh, or ring the enclosure with a low electric fence. You will need a special feeder (to keep pelleted food clean and dry) and a trough or low bucket for water.

FEEDING

Apart from household or garden greens, laying hens will need grain, grit and a ready supply of either meal or layers' pellets. Chickens love fresh greens. When the cabbages have become a bit leggy, or a vegetable marrow is oversized, suspend them from a fence pole or similar, so that they are hanging at about beak height and the birds can have a good feed without stomping the greens into the ground.

If you want to increase egg production, add a little cod-liver oil to the diet. Figures suggest that just a small amount of oil boosts egg production, increases general vitality, and goes a long way to building up resistance to disease. As a rough guide, one teaspoon of oil per day is enough for a dozen or so small chicks, while adult bird needs about 1 oz to every 6–8 lb of mash. All that said, a good ready-mix food will almost certainly contain such a supplement. Grit is essential to all poultry – in their gizzard to help them grind up food, and of course as a shell-making material. Though traditionally most egg producers favored using crushed flint, they now

147

use crushed limestone or crushed shell. When it comes to feeding, beginners are often confused about the term 'mash.' The problem is that the term is used to describe both a balanced mixture of meal as might be purchased from a supplier – 'Baby Chick Mash,' 'Layers Mash' and so on – and a home-made mix of feeding stuff, such as a mash made from a mix of cooked potato peelings and oats. When it comes to mixing a wet mash (a home-made mash and any other feed that needs to be mixed with water) it is important

that it is mixed to a moist crumbly consistency. On no account should it be sloppy like thin porridge. When it comes to feeding found and wild foods, the biggest dangers are salt, rhubarb leaves, green sprouted potatoes and deadly nightshade. The last three are simply poisonous. The salt is a little more tricky, because, while they need it in their diet, too much is a killer. Salt is already in various ready-mix feeds, so all you have to do is to make sure that you remove salt from plate scrapings before you empty them in the chicken run.

TROUBLESHOOTING

SUNLIGHT Fresh air and sunlight are vital if you want the birds to stay healthy.

VERMIN Rats and mice are attracted by the food that you put down for the chickens. Don't put down more food than the birds can eat, keep moving the feeders around, and get a cat.

MITES (LICE) These are like little specks of red or gray dust. Cleanliness is the best prevention. If egg production is going down, or you see the birds scratching, rubbing and looking a bit uncomfortable, suspect lice or mites and use an appropriate organic dusting powder.

SCALY LEG This is an unsightly and crippling condition caused by a parasite. If you see a bird limping, or think that a leg looks deformed, suspect scaly leg and use one of the appropriate treatments.

ABNORMALLY HEAVY Common problem in small flocks where birds are fed too much bread or potatoes. In extreme cases the bird will have a rock-hard swollen abdomen. Treat by throwing corn into the grass so that the bird is forced to scratch and exercise.

GAPING The bird stands still with the neck stretched and the mouth opening as if to cough or choke. This indicates gape worms in the throat. Sometimes the condition is so advanced that the worms can actually be seen at the back of the throat. Traditionally this condition was treated by hooking the worms out with a feather and painting the back of the throat with paraffin or castor oil. The condition indicates that the ground is contaminated. Best treatment is to move the birds to another plot.

SUDDEN COLLAPSE The bird suddenly goes limp and starts to breathe with small shallow gasps. The condition indicates some sort of catastrophic rupture of a blood vessel or organ. The condition can be brought upon by heat exhaustion or fright, such as that caused by a loud noise or a low-flying aircraft. There is not much you can do other than keep the bird cool and quiet.

QUESTIONS AND ANSWERS

• **What are bantams?** Bantams are a race or type of miniature fowl, usually about half the size of common birds. Bantams are good for option for self-sufficiency in that they are small and vigorous. As an average a bantam will lay 100-150 small eggs per year.

• **Why do chickens peck each other?** Though chickens tend to peck each other when they are variously bored, underfed or overcrowded, just occasionally there are chickens that peck for the pleasure of it. All you can do is separate them from the rest of the flock.

• **How can I get a broody hen?** Some breeds are naturally broodier than others – usually, but not always, the heavy breeds. If you want a few broody hens to sit on eggs, get a breed that will readily go broody, like a Silkie. Geese also make good sitters. Broodiness can be encouraged either by leaving eggs in one of the nests, or better by leaving pot-eggs in the nest and ensuring that the house is warm and dry. A broody hen will sit on the nest for long periods, and generally ruffle and spread her feathers when approached.

• **Is it better to buy chicks or pullets?** Beginners should get eight-week-old pullets rather than chicks; then you will not have the cost of special heating and chick food. Make sure that you get your birds from a well-known breeder. Once you have more experience, it is a good option to breed your own stock.

• **How do I rear chicks under a broody hen?** If you already have your broody hen, your next requirement is a warm, waterproof coop – a cube of about 2 ft. Position the coop in a warm, sheltered position complete with a water bottle and a flat feeding trough or board. When the broody hen is established and comfortable, wait until dusk, and then take your chicks (all dry and fluffy) and present them to the hen. When the hen and chicks are settled, give the chicks a feed of chick food/corn and milk. Feed the hen her usual fare of grain, mash or pellets. Repeat this procedure every 2–3 hours or so from dawn until dusk. Continue for about 7–8 weeks, gradually enlarging the enclosure or run.

• **What is dry mash?** Dry mash is a carefully compounded mix that contains all the necessary nutrients.

• **Why is there blood on the eggs?** Do not worry if a pullet's first few eggs are smeared with a few drops of blood – it is quite usual. If the condition persists, the likelihood is that the bird is injured in some way or other.

• **Is it true that kitchen waste can affect the taste and smell of an egg?** While it is possible for kitchen scraps in the feed (like onions and garlic) to taint an egg, the chances are that the egg has either absorbed the smell/taint by being in a confined space with onions, or by actually touching them.

• **Do I need to feed chicks that have just hatched?** There is no need to feed the chicks until they are about two days old because they will have already taken enough nourishment from the egg to tide them over.

• **Can chickens eat wild foods?** Much depends upon the food, but they will eat everything from elderberries and blackberries through to chestnuts, conkers, crab apples and acorns. Be wary about boiling up, say, acorns and presenting it to them as a feed; it is much better to let them scratch for what they need.

KEEPING DUCKS

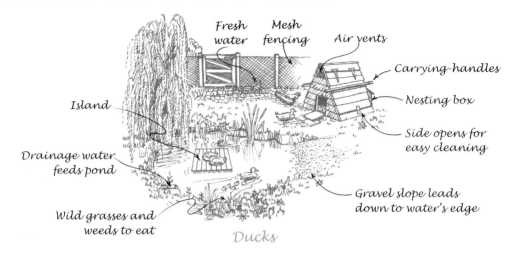

Fresh water

Mesh fencing

Air vents

Carrying-handles

Nesting box

Side opens for easy cleaning

Island

Drainage water feeds pond

Gravel slope leads down to water's edge

Wild grasses and weeds to eat

Ducks

Why keep ducks? The answer is beautifully simple: ducks eat kitchen scraps, they lay eggs, they are tasty, and above all they are good fun. The only down side is obvious – ducks need water. A large lake would be good, but they can get by with a small pond, a stream or even a collection of water troughs. In addition to kitchen scraps, and grit or shell for making eggs, they need pellets and a little fish oil in their feed to keep them in fine feather.

BREEDS

AYLESBURY **Plump, white-feathered bird that produces large, white-blue-green eggs that have an excellent taste.**
KHAKI CAMPBELL **Khaki-feathered bird that is one of the best laying breeds, and fair for eating. A mature bird will lay up to 300 white eggs a year.**
INDIAN RUNNER **Good choice for rough ground, if you want to see your ducks running wild. Fine for eggs, but a bit stringy for eating.**

CHOOSING DUCKS

Decide how much space and time you have to devote to ducks, whether you want eggs, meat or both, and then choose a suitable breed.

HOUSING

The ideal duck house is small, cosy, mobile and easy to clean, like the traditional A-framed ark. If the ground is looking a bit stale, simply pick the ark up by the integral carrying-handles and swiftly move it to a fresh location. The area needs to be ringed with close-mesh chicken wire. Allow about 19–24 sq yd per bird. You also need a feeder, and a trough or low bucket for water. If predators (such as foxes) are active in your area, cover the enclosure with a ceiling of wire mesh, or ring it with a double fence and a low electric fence.

FEEDING

Apart from kitchen scraps, ducks need grit, and either meal or layers' pellets, and a supply of fresh greens (see Chickens, page 147).

KEEPING GEESE

Geese are very nearly self-sufficient. Given a good-sized patch (say 24 sq yd per bird), they can manage for the best part of the year without extra feeding. They also eat kitchen scraps, lay eggs and are very tasty. The down side to keeping geese is that they are noisy, they make a mess, and they can be aggressive.

CHOOSING GEESE

Stand back and look at what you have, and decide what you need. For example, you might have about ½ acre of overgrown orchard that needs attention, and you would like the eggs. You could mow the grass, but perhaps better still you could fence it and stock it with geese. The geese would get as much food as they could manage, you would get huge eggs, and your orchard would look all the better for it, with short-cropped grass, not so many weeds, and all the bugs eaten.

Nesting box

Good ventilation

Geese

HOUSING

Geese can be housed just about anywhere – an old shed, a large crate with a bit of a roof, an old hen hut or lean-to, a shed with sod or straw bale walls wrapped around with mesh and a tin roof. Anything will serve, as long as it is dry, high enough for the geese to stand up in, with a space to lay eggs, and fox-proof. The area needs to be ringed with chicken wire. You also need a feeder trough, and a trough or low bucket for water. As for foxes, the geese need to be locked in at night, but a flock of six or so geese can be more than a match for a fox.

BREEDS

EMBDEN Pure, glossy, white bird that is a good layer, good to eat, very hardy and easy to keep; it weighs in at about 18 lb.

TOULOUSE Gray- to white-feathered bird that can give up to 60 eggs a year, and is good to eat; average weight is about 30 lb.

FEEDING

Geese can survive with not much more than free-roam grazing, kitchen scraps, garden greens, grit and a ready supply of water; but, if you are trying to fatten them up or get them to produce lots of eggs, they might need extra feeds of cereal or whatever is available (see Chickens, page 147).

TROUBLESHOOTING

Not enough space While geese are not too fussy about housing, they do not like to be confined in too small a space. As well as the fenced area, they also need free grazing.

Mixing breeds Different breeds of geese must be kept in separate yards and houses because they do not do well when they are mixed together.

KEEPING SHEEP

Clean tail area

Bright eyes

Even growth to fleece

Good set of lower teeth

A healthy sheep

Why keep sheep? This is the question we asked ourselves every time one of our ewes got stuck in the fence, in the hedge, in the ditch, under the Land Rover, upside down in a dip between two furrows, in the chicken shed – if there was a place to get stuck, then one of our sheep would manage it. Then there was the dreaded blowfly, bloat, nostril fly, and all the other flukes, rots, pests, parasites, bugs, sicknesses, infections, maladies and diseases that wanted to munch on the sheep.

Despite all that, however, sheep do have their good points: they are relatively low in cost, they provide meat, they can live on rough land, they can be easily housed, they are docile, they will mow grass for free, the lambs look beautiful, and – best of all from our point of view – they supply wool. A six-ewe flock will not only cover costs, but will also give you a steady supply of legs for roasting, chops for grilling, meat for stewing, and of course wool.

QUESTIONS TO ASK YOURSELF

- Do I want meat, or wool, or both?
- At about 3–4 ewes to 1 acre, depending on the quality of the grass and allowing for, say, a couple of lambs for each ewe at lambing time, how many sheep do I want to run?

- Do I want a ram, or am I going to run my ewes with a neighboring farmer's sheep?
- Have I got enough land to fence it off into little paddocks?
- Am I aiming for a simple spring-summer-autumn-winter approach (lambing in spring, fattening lambs in summer, killing them in autumn, eating them in winter), or am I prepared for the expense of overwintering indoors?

CHOOSING SHEEP

The trick here is to choose a breed to suit your area and your needs – lots of wool, heavy weight, small size, and so on. For example, when we lived in Leicestershire, we chose a breed called Leicestershire Blackface, for three good reasons: they were bred to suit local conditions (the weather and the shape of the land); they were readily available; and, most important of all, the local farmer knew them inside out. Although there are perhaps hundreds of breeds, there are three main classes of sheep – short-woolled, long-woolled and mountain sheep. The following list will give you just a flavor of the possibilities.

BREEDS

LONG-WOOL SHEEP include **Devon Long-wool, Leicestershire, Border Leicestershire, Wensleydale, Cotswolds, Hampshire Down and Lincoln.**

SHORT-WOOL SHEEP include **South Down, Suffolk and Dorset.**

MOUNTAIN SHEEP include **Jacobs, Welsh Mountain, Cheviot and Dartmoor.**

BREEDING FACTS

- Sheep reach sexual maturity in the second year after birth.
- Ewes mated in autumn of one year will be heavily pregnant 140–150 days later in the spring of the following year.
- 'Raddling' is the term used for the procedure of fitting and using a color-carrying harness that shows when the ewes have been mated by the ram. By changing the colors on a certain date, you can begin counting days.
- Just before lambing, the ewe will lie down and start groaning and straining. If all goes well, the water-bag appears first, followed by a nose and two feet, and then the head and shoulders; finally the lamb will plop onto the floor.
- Once the lamb is out, the mother will lick all the mucus away from the lamb's face.
- If the water-bag is followed by the back feet, or by a back foot and a nose, for example, you should call in the help of a friendly farmer or a vet, as pre-arranged.
- The lambs will suckle within a couple of hours of being born.
- If a lamb is rejected, it will need to be fostered either by you or by another ewe.
- Lambs can be bottle-fed with formula milk, just like a baby. Bottle-feeding is a day-in, day-out task.
- Lambing once started must continue to its end. Always be fully prepared.

HOUSING

For the most part, sheep do not need housing, apart from a rough shelter at lambing time. A good idea is to build a temporary shelter using square bales of straw and corrugated iron sheet, with the whole shelter gated and fenced at the front, and strapped up with rope.

Feeding orphaned lambs

Straw bales used as shelter

Patent lamb feeder

Straw bedding

FEEDING

Most sheep are 'grass sheep,' meaning that, depending on the breed, they can thrive on second-class or even third-class pasture. You will need to give extra food to ewes just before and just after lambing (corn, chopped rutabagas or concentrated pellets) and to lambs you are trying to fatten (rutabagas, hay or linseed cake). As for winter feeding, if you have, say, six ewes, and you aim to fatten and kill the lambs before winter, the ewes will need one bale of hay a day for basic winter feed, plus a rising intake of grain and supplements as winter runs into spring and lambing time.

SHEARING AND SPINNING

Shearing and spinning are crafts that are best learnt by watching an expert at work. In a hot summer the fleece can be removed simply by easing it away from the body. Spinning is best done on a traditional large horizontal wheel.

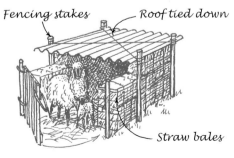

Fencing stakes   Roof tied down

Straw bales

A temporary shelter

KEEPING GOATS

WHY GOATS?

Goats are more manageable than cows, are inexpensive to buy and feed, can survive just about anywhere, are less choosy about food (they browse rather than graze), can give about 6–8 pints of nutritious milk a day (perfect for invalids and for people who are allergic to cows' milk) and generally fit very nicely into a small-scale self-sufficient set-up. On the other hand, they will, if given half a chance, strip your food garden bare, find their way over, under or through a less than perfect fence, and need milking once or twice a day. All in all, they are a good option if you want milk, and if your land tends toward being mixed scrub rather than rich meadow.

BREEDS

Every goat breed has its own qualities – small in size, fine coat, hardy, high milk yield, and so on. Of course, within any breed there are also going to be good and bad examples. The best thing to do if you are a beginner is arm yourself with our general guide, make contact with one or other of the goat societies, and take it from there.

WARNING If you are worried about spiky horns and small children being a dangerous mix – and most but not all breeders agree that they can be – choose a goat that has been polled or disbudded.

BUYING A GOAT

Once you have made contact with a recommended registered breeder, and when you have a clear understanding about set-up costs, housing, feeding, milking, how much time is involved, and all the other considerations,

GOOD MILKING BREEDS

ANGORA A good choice on two counts – they are very good milkers, and there is a ready market for their fleece or mohair.

ANGLO NUBIAN This is the breed to choose if you are looking for a high milk yield and want to make cheese. The milk is high in butter fat.

TOGGENBURG This is a very popular breed, much favored by beginners who are looking to get a good milk yield at a relatively low feed cost.

SAANEN A good all-rounder, with a short fine coat. They are good for milking and have a placid temperament.

then comes the wonderfully exciting business of actually choosing and transporting the goat. Regarding whether you should go for a mature goat that is in milk or a younger one, much depends on your needs and your patience. The best all-round advice is to start out with a couple of doe kids that have just been weaned, and then gradually experience

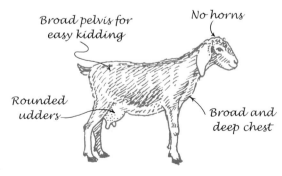

Broad pelvis for easy kidding

No horns

Rounded udders

Broad and deep chest

Angora goat

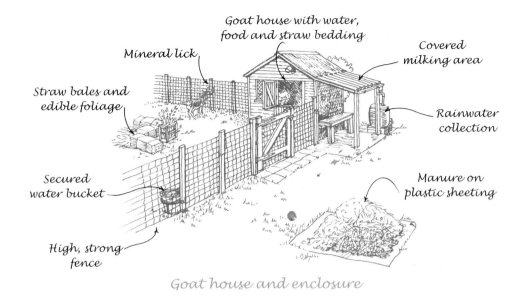

Goat house with water, food and straw bedding

Mineral lick

Covered milking area

Straw bales and edible foliage

Rainwater collection

Secured water bucket

Manure on plastic sheeting

High, strong fence

Goat house and enclosure

all the growing and breeding stages. In this way, you and the goats will grow up together and become friends, and as you go you will begin to appreciate just what it is that makes a goat tick, all before you start milking.

As for the tricky business of transporting, it is best either to fetch them in a closed van or horsebox, or to let the breeder arrange the transport. Just remember that goats are agile and very intelligent.

HOUSING

Though much depends upon how much ground you have and how many goats you want to keep, the ideal is to house the goats in some sort of weatherproof shed that is tall enough for you and the goats to move around in, to have a fenced and concreted area around the shed, a clean covered area just outside the shed for milking, and access to scrubland for browsing. If you want to keep the goats in the enclosure and bring in food, then the enclosure needs to be that much bigger. Lots of space equals happy contented goats which in turn equal a good milk yield.

A good idea, when planning your particular set-up, is to imagine yourself dealing with the goats on a day-to-day basis, and plan and build the housing accordingly. In the normal course of events, you would get up in the morning and look at the goats, and maybe let them out if the weather is right. You will have to fetch water for them, give them additional feed, clear away muck and move it to a dung heap, walk them to their milking area, wash your hands and do the milking at a comfortable working height, and so on. Look at your proposed location, take notes about the water supply, the distance from the house, and other details, and then build the layout to suit.

Most experts reckon that a well-posted, well-stretched wire-mesh fence at about 4–5 ft high is good for around the compound, and an electric wire-strand fence is good for around the field. Goats will generally stay put as long as they are well fed, not too bored, and not tempted by forbidden delights such as fruit trees and vegetable patches. There is always going to be a difficult goat, but that is half the fun of it.

155

FEEDING

If your goats are going to be kept permanently indoors, there needs to be an unlimited supply of water and hay, a salt lick and an additional supply of either a concentrated goat feed or a mix of something like crushed oats and flaked maize. If you are going to let them browse on a piece of scrubland, hold back with hay until the weather gets cold and then give them about 12 lb of hay a day. If they are giving milk, supplement the hay with around 2 lb of pellets.

BREEDING

The female goat is ready for mating at about 18 months old, in the autumn, when the vulva shows wet and red and the tail wags from side to side. The season lasts anything from a few hours to a few days and will return every 21 days or so until she is mated. As for getting a goat mated, you have the choice of keeping your own male, using a stud goat, or using artificial insemination. The easiest option in the first instance is to take your goat to stud. Make sure that you use a male of the same breed, and one with a good pedigree and a proven track record. The gestation period (the time from mating to birth) is usually about 150 days, give or take 7 days.

As the time for kidding approaches, the udder will start to grow and fill. The vulva will become puffy and show a discharge, and the goat will become unsettled and noisy. The labor starts with a lot of straining, and very soon results in the expulsion of the kid nicely wrapped in a membrane bag – front hooves first, followed by the nose and then the rest of the body. With warm water, soap, paper towels, scissors and antibiotic spray/wash at the ready, first use a paper towel to clear the kid's mouths and nostrils. Once the afterbirth

has been discharged, the birth is over. Things can always go wrong – for example, a smelly discharge, a long labor, the head showing first, a swollen udder, the kid or the mother looking very distressed – so make sure that you have an experienced friend at hand, and the vet's phone number at the ready.

MILKING

Leave the mother with the kid(s) for about two weeks, to give them a fair start, and then take them off and start milking. Be aware that, while many goat-keepers milk twice a day, there are others who reckon that it is best to milk just once a day. Certainly, the once-a-day method will give you less milk, but the time spent milking is cut by half. Ask yourself whether you want more time or more milk.

When you are ready to milk, put the goat on a leash and lead her to the clean area that you have set aside for milking. Set the goat on the low table (the one that you have built so that everything is at a comfortable working

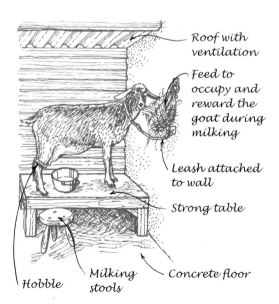

Roof with ventilation

Feed to occupy and reward the goat during milking

Leash attached to wall

Strong table

Concrete floor

Hobble

Milking stools

Milking area

height), clip the leash to the wall/stall and fit a hobble on the back legs so that she cannot kick the milk basin over. When we were beginners, we had a goat who was quite a character, and who took pleasure in kicking the milk bowl over. A hobble is a strap that restrains a leg, making milking easier.

Start by washing the udder with warm water. This removes dust and dirt and encourages the goat to 'let her milk down.' This done, encircle one of the two teats with thumb and hand so that the thumb is at the front and the open palm at the back, and gently squeeze with the thumb so that the milk in the lower part of the teat is contained and under pressure. Now apply a firm but gentle downward stroking pressure so that the milk squirts out. Clean and clear the teat by wasting the first squirt onto the ground, and then ease back with your thumb so that more milk flows into the teat. Repeat the thumb pressing and stroking procedure as described, but this time direct the milk into your pail or bowl. When one side of the udder is empty, milk the other teat in the same way.

When the job is done, walk the goat away from the milking area. Finally, take the milk indoors and strain it through several layers of clean white cotton cloth or filter papers, and store it in the fridge ready for use.

MILKING QUESTIONS AND ANSWERS

- **Does goat milk taste funny?** Not a bit of it. I don't like it warm, so I chill it before drinking, but all in all I would say that it tastes creamy with a slightly nutty background flavor.
- **Can I freeze the milk?** Yes. Pour the milk into lockable freezer bags and stack them in the freezer.

- **How long is the milking cycle?** From the birth of the kid(s), reckon on two weeks for giving the milk to the kids, and about 6–7 months' milking, with the yield peaking at about 8–9 weeks. This gives the goat a rest for about 4 months or so.
- **Can I miss a day's milking?** Yes, you can just about miss a couple of days if the kids are at hand; you simply bed them in with mum and let them drink her dry.
- **Can the doe make milk without having kids?** No, the female goat is just like any other mammal in this respect – she must have babies.
- **How can we dry the doe off?** Cut back on the additional feed. Start by cutting back to milking once a day, and then every other day, and so on until her milk-making system shuts down.

MAKING CHEESE

Cheese making (see pages 174–175) is one of those almost magical activities that is surrounded in mystery … one day you have a bucket full of milk and a week or two later you have a nutritious food that will, under the right conditions, last for months. The best advice for beginners is to start by making a basic soft cheese, and to try more complex hard types when you have some idea of your skills and needs.

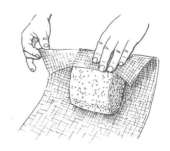

Wrapping the cheese

KEEPING A COW

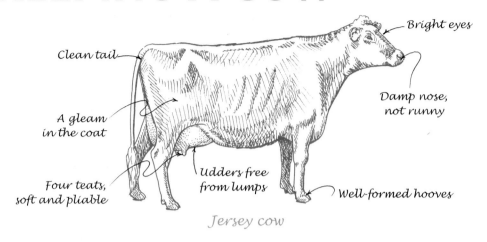

Bright eyes

Clean tail

Damp nose, not runny

A gleam in the coat

Four teats, soft and pliable

Udders free from lumps

Well-formed hooves

Jersey cow

We bought our cow, Doris, at a farm auction for not much more than a bag of beans – a bargain because no one else wanted a Jersey cow with only three teats. When she stood there with her enormous, sad eyes and her three huge teats leaking milk, we could not resist her and made a bid. The good news is that Doris was easy to milk, and generally a great character. She only gave us 17 pints of milk a day, but it was more than enough for us and our two boys, and there was some left over to make into butter. Doris was just right for us since we did not want vast quantities of thin, poor-grade milk, and as we milked by hand it did not matter to us that she did not fit in the great four-teat, machine-milking scheme of things. As for her calves, while the heifers sold well, we could barely give the bull calves away.

FACTS, FIGURES AND QUESTIONS

- If you are planning to make your own hay, and if we take it that one cow needs about 1 ton of hay for winter feed, then it follows that you need about 2 acres of grass – half for the cow and half for making hay.
- If you are going to buy in the hay, where are you going to store it?
- What are you going to do with the excess milk? Are you going to feed it to any calves, make butter and cheese, or feed it to pigs?
- When it comes to milking, are you going to milk twice a day (TAD) and get maximum yield, or once a day (OAD) and a much lower yield?
- Are you up to the task of milking every day of the year?

Finger and thumb trap the milk in the teat; then the other fingers squeeze the milk out

Milking

- If you are going to make butter, cheese and/or yogurt, are you kitted out with the necessary equipment (see pages 172–175)?
- Your cow will need some sort of shelter to keep off wind, rain and sun.
- You will need a shelter for milking.

BREEDS

DEXTER Small; black or red coat; becoming rare; good to milk and good to eat; high-yielding.

FRIESIAN Large; black and white coat; very high milk yield; good to eat.

JERSEY Medium-sized; light brown coat; makes a good 'house' cow; medium yield of rich cream milk; good temperament; easy to handle; needs housing in winter.

CHOOSING A COW

When it comes to buying a cow, there are lots of potential problems. For example, you will not be wanting a freshly calved heifer, because you would both be milk virgins – the first time she has been milked, and the first time you have done the milking. It follows that the ideal cow is one on her second calf that has been hand-milked. Ask yourself why she is being sold. Are her udders in good order? How much milk does she give? If you buy a cow plus her calf, you will suddenly have two animals to look after. There are also a number of breeds to choose from.

BREEDING FACTS

- Heifers are ready to be put to the bull at about 18–24 months.
- The gestation period – time from mating to calving – is about nine months.

- If you are aiming for, say, a spring birth, you need to have your cow mated sometime around the previous June.
- A cow can only be mated when she is in heat; this is shown by a discharge and by her skittish behavior.
- You have a choice: you can let your cow run with a neighboring farmer's bull, in which case the bull will know when your cow is ready; or you can have her artificially inseminated, in which case you will have to judge when she is ready.
- From about eight weeks before calving, you must increase the cow's feed – oats, hay, a nice mix of whatever is going.
- When the cow's time is near, her udders will fill and her back end will go loose and slack.
- If all goes well, the calf will be presented in much the same way as a sheep or goat: the water bag will appear, followed closely by a nose and two feet, and then the head and shoulders, and finally the whole body will come out in a rush.
- Once the calf is out, the mother will start licking it into shape.
- The best you can do is be prepared: read books, talk to local farmers, help local farmers at calving time, and have at hand any relevant telephone numbers if help is needed.

WEANING

You have a choice: you can let the calf have all the milk; you can milk the cow and give half to the calf; you can share the milking – two teats for you and two for the calf; you can let the calf feed for a week and then take it away completely; and so on. You must first sort out your needs in this area and then act accordingly.

KEEPING PIGS

Keeping pigs is a challenge; it is a good way of producing choice organic food, and it makes good money sense. The only down side is that pigs are friendly and very intelligent, which makes it all the harder when the time comes to slaughter them.

FACTS, FIGURES AND QUESTIONS

- The best option, in the first instance, is to fatten a couple of weaners up using excess milk, kitchen and garden scraps, and then maybe go on to breeding when you can better judge your skills and needs.
- Always get two pigs, as they will be happier, and consequently easier to fatten.
- You can keep pigs indoors in a barn or shed, but it is much more fun if you keep them outdoors; they can enjoy being pigs, and you can enjoy watching them.

- The easiest option when it comes to fencing is to get one of the electric fencing systems, which are easy to install and nicely portable, and they really do work.
- A couple of pigs are just the thing for clearing a patch of ground that is overgrown with weeds.
- Pigs will eat all the kitchen stuff that usually goes on the compost heap.
- Your pigs will need some sort of shelter to keep off wind, rain and sun.

CHOOSING A PIG

When choosing weaners, go for one of the traditional breeds (see opposite). You can look for a couple of orphaned piglets – they will not cost much, but you will have to hand-rear them with a bottle; or you can get a couple of 8–9-week-old weaners – more expensive, but healthier and less of a challenge.

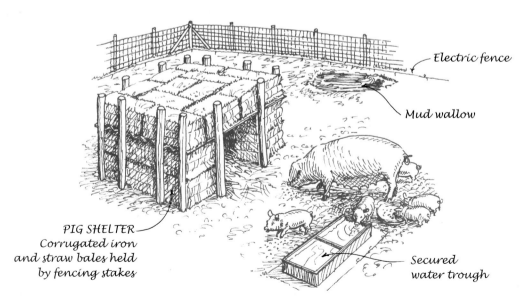

Electric fence

Mud wallow

PIG SHELTER
Corrugated iron
and straw bales held
by fencing stakes

Secured
water trough

A pig enclosure

BREEDS

LARGE BLACK **Large; black-haired; traditional lop-eared; good temperament; ideal for keeping outdoors.**
SADDLEBACK **One of two saddleback breeds, the other being the Wessex Saddleback; black with a white band; hardy; very good for outdoors.**
TAMWORTH **Good-sized; sandy or reddish; long snout; good for outdoors.**

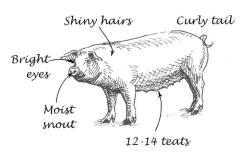

Shiny hairs   Curly tail

Bright eyes

Moist snout

12-14 teats

Landrace pig

HOUSING

A couple of pigs can be housed in an A-frame hut that has been knocked together from timber and corrugated iron, hay bales topped with sheets of corrugated iron, or indeed anything as long as it will keep off the sun, wind and rain. If your plot is particularly exposed to cold winds, then the pigs will do much better if you build a straw bale and plastic sheet buffer on the windward side, with, say, the straw being packed against the wall and the roof.

You willl soon learn, however, that one of the great pleasures of keeping pigs involves you working out how best to combat the pigs' seemingly overwhelming desire to variously eat, squash, batter and flatten their housing at every opportunity.

FEEDING FACTS

- Animals must be fed at regular intervals, because pigs seem to enjoy routine.
- They should be given as much as they can eat in 20 minutes.
- They need about 2 pints of water for every 1 lb of feed – as much as they want to drink.
- Weaners need three good feeds a day, plus all they can find.
- When the pigs are out to grass, you can cut down on the supplementary feeds.
- Pigs are fed basic meals up until about 16 weeks, and then fed to fatten.
- At 18–20 weeks, the pigs can be allowed to make real pigs of themselves.
- Pigs do well on wild food such as acorns, sweet chestnuts and elderberries.

TROUBLESHOOTING

Pig fights If you put two strange pigs together, the chances are they will fight. While this is not too much of a problem, because it generally does not last too long, the danger is that they will hurt themselves and/or damage the housing. If you do have to put strange pigs together, do it when it is nearly dark.

Grooming It is a good idea to groom the pigs with a stiff brush; they seem to enjoy it, they look better for it, and you can check them over for problems while you are doing it.

Sudden death If your pig dies suddenly for no good reason, you are required by law to make contact with a vet, just in case it has one of the serious diseases - anthrax, foot-and-mouth, rabies or swine fever.

Poisoning This shows as diarrhea, loss of appetite, thirst and sometimes convulsions. Possible causes are too much salt or poisonous weeds. Avoid giving pigs too much salty kitchen waste. Watch out for hemlock, yew and privet.

KEEPING BEES

Why should you keep bees? They are fascinating and challenging, and at the end of it all you get honey.

FACTS, FIGURES AND QUESTIONS

- Sooner or later, you will get stung. You must have an allergy test to make sure that you are not hypersensitive.
- Beekeeping will, if you are averagely lucky, pay for itself, keep you in honey, and maybe bring in a little money.
- Contact a beekeeping association for advice, set-up stock and secondhand equipment.
- In the knowledge that bees will collect nectar from a 2 mile radius of the hive, a bit of research on your part will tell what sort of honey you can expect – clover, rape or lavender, for example.
- How much work will I need to put in? We know two beekeepers: one says that a couple of long days every year is enough; the other says that one inspection every two weeks from mid-spring until late summer does the trick.
- How many hives does a beginner need? Most experts reckon that one hive is a bad idea, two hives are better, and three are perfect.

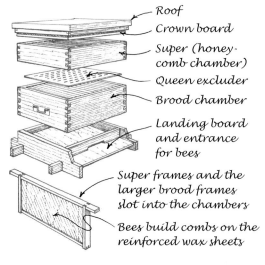

A hive

Roof
Crown board
Super (honey-comb chamber)
Queen excluder
Brood chamber
Landing board and entrance for bees
Super frames and the larger brood frames slot into the chambers
Bees build combs on the reinforced wax sheets

- The hive consists of a base or stand, a bottom board, two brood chambers, a queen excluder, a couple of supers, an inner cover and a metal top.
- Apart from the hive itself, a beginner needs a protective outfit (consisting of combined mask, hat or helmet and veil, a cover-all suit and gloves), a scrape/hive tool and a smoker.

GETTING YOUR BEES

You can set yourself up with an empty hive, and go hunting for swarms, and then just sit back and hope it works out, but the easiest way of getting a colony is to purchase them in spring from a recommended, established breeder. You can buy them by weight, or by the frame, or by the hive. Good advice for a beginner is to get a starter pack: a new hive made up with a laying queen and bees, plus three supers (honeycomb chambers) made up ready for honey production. Look around for a

DRONE
QUEEN
WORKER

Bees

well-established breeder, make a phone call to place an order, turn up with a suitably sized vehicle and take delivery.

EXTRACTING HONEY

1 Get yourself ready with all your equipment: bee suit with veil, gloves, smoker, hive tool, fume pad, bee brush, barrow or trolley and bee repellent.

2 Lightly smoke the bees and remove the top from the hive. Use the tool to loosen up the main hive bodies.

3 Pour a minute amount of repellent onto the fume pad and put it on the super, until the bees are gone.

4 Swiftly lift the frames from the hive and place them in a spare super on the trolley; continue until all frames are on the trolley.

5 Replace the hive top and take the fume pad well away from the hive.

6 Wheel the trolley to your honey room.

7 Take the knife, slice through either side of each frame to remove the wax caps, and put the frame in the extractor.

8 Set the extractor in motion so that the honey runs free into a bucket, first one side of the frame and then the other.

9 Having captured all the honey, go to the kitchen and run it through a sieve.

10 Warm the room up to keep the honey moving, and run it into jars.

11 Label the jars and wash up.

BRIEF BEEHIVE GLOSSARY

BASE The base on which the hive sits.

BEE SPACE A ¼–⅜ in. space between component parts of the hive.

BOTTOM BOARD The floor of a hive.

BRACE COMB A component part that links two combs.

BROOD CHAMBER The part of the hive in which the brood is reared.

CELL One of the hexagonal chambers in a honeycomb.

COMB FOUNDATION A man-made structure made up of thin sheets of beeswax to give the bees a start.

DECOY HIVE An empty hive use to attract stray swarms.

FRAME Four pieces of wood designed to hold a honeycomb.

HIVE BODY A wooden box that holds and encloses the frames.

LANGSTROTH EQUIPMENT A term used to describe standard-sized equipment, named after the Reverend Langstroth (c. 1850).

QUEEN EXCLUDER A device that restricts the drones and queens to certain parts of the hive.

SMOKER A device with bellows and nozzle, used to subdue bees by puffing smoke at them.

SUPER Any part of the hive used for storing honey.

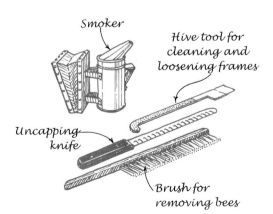

Smoker

Hive tool for cleaning and loosening frames

Uncapping knife

Brush for removing bees

Extracting honey

THE PANTRY

DRY-CURING BACON

Dry-curing – sometimes called just 'curing' or even 'salting' – is a simple process that involves burying a slab of belly pork in salt for 7–9 days and then hanging it up in a cool, airy place to dry for 9–20 months. The advantages of home-cured bacon are that the joint is firm and hard to the touch, it cuts like a very hard cheese (so much so that a cut slice comes away from the joint in a wafer-thin, stiff curl) and it emerges dry, crisp and full of flavor when it is fried in the pan.

HOW TO TURN BELLY PORK INTO BACON

You will need:
- Two large slabs of boned-out belly pork, cut from the same pig – the thickest ones you can find.
- Large, shallow, waterproof, wooden washtub, or shallow, plastic storage box, long enough to take the pork when it is laid down flat. If you use a washtub-type box, make sure it is free from metal – no nails, screws or fixings.
- 50 lb bag of salt.
- Eggcupful of saltpeter.
- Half a cup of brown sugar.
- Black pepper to taste.
- Enough fine-weave muslin to double-wrap both slabs of pork separately.
- Large, bodkin-type needle and strong linen thread.

1 In late autumn, take the pork (five days after it has been killed) and trim off all loose ends and scraggy bits. If the pork is yours and completely organic, so much the better.

2 You and any helper(s) should remove all metal items, such as watches, rings and bangles, roll up your sleeves, wash your hands with fragrance-free soap, and clear the surface ready for action.

3 Pour a thick layer of salt into the bottom of the box or tub and place one pork slab on top of it, skin-side down.

4 Take the saltpeter and rub a pinch of it here and there into the uppermost surface of the meat. Try to work it into anything that looks bloody. Place the second slab on top of the first and repeat the procedure.

Rubbing saltpeter into the belly pork

5 Heap more salt, as well as the sugar and pepper, over the pork and rub it into every crack and cranny. Continue until every part of the pork has been worked over in this way.

Rubbing in salt

6 Remove the two slabs from the box. Then put one slab skin-side down in the box on a bed of salt, cover it with salt, and set the second belly on top of the first, skin-side up, so that you have a meat-salt-meat sandwich. Bury it all in salt, and leave it in a cool place overnight.

7 Next morning take the box, pour off the pools of liquid, make sure there is plenty of salt between the two slabs, and once again bury the whole thing in salt. Repeat this procedure every day, always making sure that the slabs are completely covered in salt. Note – do not pour the liquid down the drain; it is best to bury it.

8 After about a week or so, when there is no more liquid to pour off, remove the slabs, brush away the salt, loosely wrap them in the muslin, and hang them up in a cool, draughty place to air-dry.

9 Check them daily, keep them cool, and leave them hanging for the best part of two weeks.

10 When the part-cured bacon feels dry to the touch, take it one slab at a time, double-wrap it in muslin, and sew it up with the needle and thread to make a parcel; it should be completely covered with the muslin. Store by hanging in a cool, dry, well ventilated place.

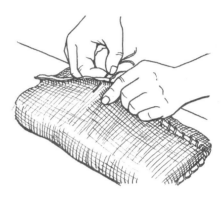

Covering with muslin

11 Finally, after at least nine months of drying, when the bacon feels rock hard, have a trial fry-up. If all is well, it should slice cleanly and firmly to produce crisp, hard rashers – absolutely beautiful!

CURING TIME

Curing pork takes 7–9 days of salting for every 1 in. thickness of meat. The drying stage should last at least nine months, but sometimes it can take up to 20 months to achieve the perfect result.

CORRECT SALTING

Signs of poor salting during the curing process include discoloration, salt crystals on the surface, and soft or dry, stringy texture. In order to avoid this happening, always take great care at the salting stage.

MAKING SAUERKRAUT

If you have never tasted sauerkraut, the very notion of eating fermented cabbage might sound a bit unpleasant, but prepare to be surprised. Sauerkraut is a winner on many counts: it is easy to make, it is high in vitamin C and low in calories, it is a good way of storing cabbage and, best of all, it makes an extremely tasty treat.

Making sauerkraut is simple – all you need to do is slice the cabbage, bed it down in salt and leave it until it ferments (see below for details). As for how much salt to use, it works out at about 2 percent by weight, or about 1 lb of salt to every 40 lb of sliced cabbage. Because salt and metal do not mix well, you should only use plastic, glass or pottery containers.

HOW TO MAKE SAUERKRAUT

You will need:

- 20 lb of sliced cabbage; this will work out at about ten or so cabbages depending on size.
- Good-sized stainless-steel knife.
- Large wooden chopping board.
- Kitchen scales.
- 12 tablespoons of pickling salt.
- Large plastic or glass bowl.
- Two or three tea-towel-sized pieces of muslin.
- Wooden spoon.
- Eleven 2 lb glass storage jars (the type with glass tops and snap lids), all well washed and dried.
- Brine (1 teaspoon of pickling salt to a cup of cold water) as needed.
- Large pan, big enough to take 3–4 jars, with a trivet in the base.

1 In autumn, when your greens are full, fat and ready to jump out of the ground, cut ten or so cabbages and remove all the outer leaves and inner core.

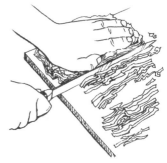

Chopping the cabbage

2 One cabbage at a time, take the knife and the board and slice the cabbage into quarters and then into thin slivers.

3 Weigh the shredded cabbage until you have 2 lb on the table.

4 Measure 2 teaspoons of salt into a saucer.

5 Layer the sliced cabbage and the salt into the bowl (cabbage, sprinkle of salt, cabbage, more salt and so on) until you have the 2 lb of cabbage nicely salted. Cover the bowl with the muslin.

Adding salt

Filling the jar

Sterilizing

6 Wait for about five minutes for the salted shreds to collapse and go limp. Use your hands and the wooden spoon to stuff the shreds into the jar, so that the jar is as full as possible.

7 Repeat this procedure for all 2 lb of cabbage and all the jars, and cover the jars with the muslin cloth.

8 Having stuffed all the jars with salted shreds, and waited for 2–3 hours for the cabbage to settle and the juice within the jars to rise, top the jars up with brine so that the cabbage is completely covered.

9 Fit the rubber rings on the lids, wipe the rims and snap the levers to hold the lids secure.

10 Store the jars in a dark, cool place and leave them for about two weeks. At the end of this time, you can either start munching, or you can sterilize the jars for long-term storage.

11 If you decide to sterilize, put the pan on the stove, sit the jars up to their necks in water in the pan, ease open the lids' snap mechanism so that any trapped gas can escape, and bring the water to a boil.

12 Simmer the jars for about 30 minutes; don't allow the jars to bang together (you may need to protect them with cloths). Snap the lids closed and let them cool.

TESTING THE SEAL

When cold, release the wire snaps and, while holding the lids, lift the jars. You have success if the seal is not broken and the lids remain in place.

CRISP FRIED POTATO CAKES AND SAUERKRAUT

Here is a traditional recipe using sauerkraut.

1 Boil two egg-sized potatoes and mash them with milk and butter, or with a little soft goat's cheese.

2 Grate three raw egg-sized potatoes.

3 Strain the grated potato, fold it in with the mashed potato and an egg, and add salt to taste.

4 Make the potato mix into little flat cakes and fast-fry them in olive oil.

5 To serve, set the crisp fried cakes on a plate, surround with strained sauerkraut, and serve with a nicely chilled, very dry white wine.

SMOKING CHEESE

Smoked cheese is a traditional food that tastes delicious and is easy to make.

HOW TO SMOKE CHEESE

You will need:

- Smoke cabinet, or an old, rusted-out, galvanized metal dustbin – one with holes in the base and a metal lid.
- Hammer and cold chisel.
- Two iron rods long enough to pass through the dustbin.
- Barbecue-type grid.
- Four or more bricks.
- Instant disposable foil dish-type barbecue.
- Apple wood or similar fruit wood (not pine). You could also use damp sawdust rather than solid wood.
- Two 2 lb blocks of cheese (we used a hard goat's cheese but it works with cheddar).
- Water spray.

I If you do not have a smoke cabinet, use

Trash can smoker

the hammer and chisel to cut the bottom out of the trash can, and to make the four holes for the rods, and sit the grid in place.

Burning apple wood

2 Choose a dry day, arrange the bricks to make a base, place the barbecue in the center of the bricks, light the barbecue, and put the apple wood on top of the coals.

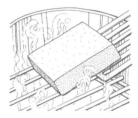

Smoking the cheese

3 Sit the trash can on the bricks and place your cheese on the grid shelf.

Tending the fire

4 Keep tending the fire, aiming for maximum smoke and minimum temperature. Smoke for 2–4 hours, depending on texture and flavor preference.

DRYING MUSHROOMS

The chances are that one day when you are searching for field mushrooms, you will suddenly be presented with an abundance. You could pickle them in vinegar, or turn them into a sauce or chutney, but the easiest option is to dry them.

Drying is a winner on several counts: it is swift, it is low-tech, and at the end of the process you have a food item that has a special character all of its own. It is very easy – all you need to do is peel off the skins, slice them up and set them over or in the range, or in a drying cabinet.

HOW TO DRY MUSHROOMS

You will need:

- Kitchen range, commercial cabinet dryer or small, home-made dryer. Our dryer was made of plywood, about 18 in. wide and deep, and 3 ft high, with muslin-covered shelves at 3 in. intervals, and an open bottom with a heater positioned just clear of the floor.
- As many field mushrooms as you can find.
- Plastic airtight containers.

1 Set up your dryer.
2 Wash, peel and slice the mushrooms.

Slicing

Drying

3 Arrange the slices on the muslin shelves and place in the dryer.

Storing

4 When the dried mushrooms feel soft and leathery to the touch, store them in the airtight containers.

Rehydrating

5 When the mushrooms are needed, take them out of the container and rehydrate them by soaking in a bowl of water until they look plump and delicious.

171

MAKING BUTTER

To make butter you need cream. There are three primary ways of separating the cream off from the milk: you can carefully skim it off by hand; you can boil the milk and then skim it off by hand; or you can use a mechanical separator. Much depends on the quantity of milk you are using, but most beginners tend to separate the cream off by hand as it is a fairly straightforward procedure.

HOW TO SEPARATE THE CREAM
You will need:
- As much milk as you want to separate.
- A piece of muslin large enough to go over the bowl.
- A large, low, lidded, wide-brimmed, stainless-steel pan.
- A stainless-steel mini-churn or jug.
- A large, flat skimming spoon.

1 Strain the milk through a muslin cloth and into the pan, and put on one side to cool for about 8–12 hours.
2 Being careful not to overly disturb the milk, take the spoon and carefully skim the cream off into the mini-churn or jug.
3 Put what is left of the milk back in a cool place for 8–12 hours, and then skim off the rest of the cream.
4 The remaining milk – now skimmed milk – can be drunk or given to the livestock.

HOW TO MAKE BUTTER
You will need:
- As much cream as you want to make into butter.
- Dairy thermometer.
- Kitchen timer.
- Largest food mixer you can find, one with a paddle blade, or a hand churn, depending on the amount.
- Piece of muslin big enough to cover one of your largest mixing bowls.
- Salt to taste.
- Large, wooden board.
- Pair of flat, wooden paddles or spoons, depending on how much butter you want to make.
- Greaseproof paper.

1 Keep the cream in the fridge or a cold pantry until it begins to go slightly sour – at least 30 hours.
2 Take your cream and pasteurize it by putting it into a pan and bringing it up to heat; hold the heat at either 160°F for 15 seconds or 145°F for 30 minutes, stirring all the time.

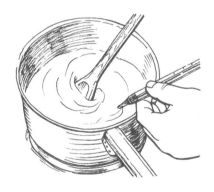

Pasteurizing

3 Remove the cream from the heat and chill in a fridge or cold room overnight.
4 Next day, pour the cream into the kitchen mixer or hand churn, leave for half an hour to become warmer than the refrigerator

Churning

7 Mix salt into the butter according to taste.

Shaping

temperature, and then set the mixer or churn in motion. Continue mixing or churning until you see and hear the cream collapse down into buttermilk and blobs of butter. The heavy, cream-plopping noise will suddenly turn to a loose, sloshing sound.

5 Continue churning until the butter becomes one large blob, and then strain it all through the muslin. Put the buttermilk to one side for kitchen use, or for feeding to chickens, pigs or family.

8 Turn the butter out on the board, and use the wooden paddles or spoons to knock it into shape.

9 Finally, turn the bricks of butter onto greaseproof paper and store it in the fridge. Larger amounts can be divided into containers and put in the freezer.

STERILIZING UTENSILS

Cleanliness is imperative in any dairy operation. To sterilize the equipment, immerse it in boiling water for five minutes, in a large stainless-steel container.

TROUBLESHOOTING

Warm or sour milk If the cream is too warm or too sour, the whole procedure will be slower. My advice is to swiftly recool the cream and then start again.

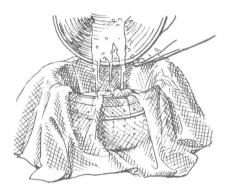

Straining

6 If it is a hot day, chill the butter in the refrigerator before you proceed. Add cold water to the butter and very slowly turn it over so as to wash out the last traces of buttermilk.

MAKING CHEESE

Most beginners tend to make soft cheese rather than hard, for two very good reasons – it is easier, and it takes less milk. For example, while a hand-sized round of soft cheese only requires 4–6 pints of milk and can be ready for eating in the space of a morning, the same-sized lump of hard cheese might take four times as much milk and then, at the end of a lengthy procedure, need to be left for a month or more. For me, making hard cheese is a very satisfying procedure.

One question is frequently asked – is there a vegetarian equivalent to rennet? The answer is yes; the vegetarian substitute is generically termed 'rennin.' Ask your supplier for details.

HOW TO MAKE HARD CHEESE

You will need:

- 2 gallons whole milk, preferably Jersey or Guernsey.
- Double saucepan large enough to hold 2 gallons. This is best obtained from a specialist supplier.
- I rennet or rennin tablet (size suitable for 2 gallons).
- Dairy thermometer.
- Large sieve with muslin to cover.
- Salt and/or herbs to taste.
- Cheese press with milk filters to fit.
- Greaseproof paper.
- Butter.

I Take your well-cooled milk from the fridge or cold room and pour it into the inner pan of the double saucepan. Heat the water in the outer pan.

2 While the water in the outer saucepan is heating, dissolve half a rennet/rennin tablet

in a quarter cup of cold water.

3 Test the milk with the thermometer; when it reaches 86°F, turn off the heat, lift the inner pan clear and put it on one side.

Adding rennet/rennin

4 Pour the dissolved rennet/rennin into the milk and stir it with a wooden spoon for a minute or so until it is thoroughly mixed.

5 Cover the milk with a muslin cloth and leave it for the best part of an hour.

6 When the milk has curdled to a rubbery firmness, take a long, stainless-steel knife and cut it into small cubes of curd.

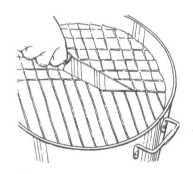

Slicing curds

7 Being very careful not to crush or squash the curds, use your hands to turn them over, so that the curds are well separated.

8 Return the pan to the double saucepan, and very slowly, over the space of an hour, heat the curds up to 102°F, turn off the heat and remove from the stove.

9 When the curds are firm, pour them through the sieve. Keep the curds, and mix the liquid whey with a little meal and give it to the chickens, pigs or dogs.

10 When the curds have cooled to about 90°F, break them into small lumps. Add a little salt, and herbs if desired, to taste.

Washing

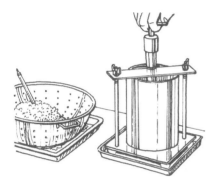

Pressing

11 When the curds have cooled down to 85°F, take the press, fit the bottom filter, top it up with curds, fit the top filter, and then either fit the weights or turn the handle, depending on the design of your press.

12 After about an hour, when the whey has stopped dripping, remove the cheese, throw away the filters, and gently wash the cheese in warm running water.

13 Carefully wrap the cheese in muslin and return it to the press.

14 After about 24 hours, ease the cheese from the press, remove the muslin, and then wash the cheese in hand-hot running water.

15 Put the cheese on a sheet of greaseproof paper in a cool, dry place.

16 Every day, inspect the cheese and turn it over.

17 At the end of two weeks, wipe the cheese rind over with a small amount of butter – just enough to give it a shine.

18 After at least a month of turning and checking, the cheese should be ready to eat. If you have too many cheeses, freeze them after they have aged.

STERILIZING UTENSILS
See page 173.

TROUBLESHOOTING
Strange taste Most problems that have to do with odd taste are caused by some sort of failure in the cleaning procedures resulting in a bacteria build-up in the cheese.

Never the same cheese twice
Regarding the question of why cheeses made from one week to the next can be so different, the answer lies in the variables – temperature, humidity, the make-up of the milk, and so on. A good option is to record details in a diary; then at least you will be able to follow the same procedures and quantities next time.

BOTTLING FRUIT

The trick with bottling – the thing that makes the difference between success and failure – is to make sure that the bottles, lids and rings are perfectly sterilized. There are four primary options: you can place the bottles in a pan as described below; you can place them in an oven; you can place them in a microwave; and you can place them in a sterilizing solution. These procedures all have their good and bad points. I think that the easiest and most reliable option is boiling and steaming, but you do have to make sure that you do not introduce bacteria. For example, it is no good going through a lengthy sterilizing procedure if at the end of it all you are going to prod and poke with an unwashed finger or spoon, or think you can save money by reusing rubber rings that are less than clean. You must be scrupulous at all times. Rings can be used twice, providing they are sterilized with the jars.

HOW TO BOTTLE FRUIT

You will need:

- As much fresh fruit as you want to bottle. I am describing how to bottle plums, but you could just as well choose apples, cherries, damsons, pears, rhubarb or whatever you wish.
- As many glass preserving jars as you can find; we use a type with glass tops, rubber rings and a wire snap-fast mechanism, but there are all manner of other Kilner-type jars.
- Salt.
- Sugar (optional).
- Deep pan or saucepan large enough to take 4–5 jars at a time, one with a loose trivet in the bottom.
- Kitchen thermometer.

1 Grade the fruit according to size, stage and ripeness. If the plums are over-ripe and/or in any way damaged, put them to one side for immediate use. It is best to use large plums.

2 Remove leaves and stalks, and wash plums thoroughly.

Preparing the fruit

3 Cut each plum in half with a sharp knife, remove the stone, check for damage, and then drop the good halves in slightly salty water to prevent the cut flesh going brown (use ½ oz of salt to 1 gallon of water).

Filling the jars

4 Wash and dry the jars, and check that the lids and rubber rings are in good order; replace any cracked rings.

5 Pack the plum halves in tightly so that the cut sides are up against the sides of the jar.

6 Fill the jars up with boiling water, or a mix of boiling water and sugar (use 8 oz of sugar to 1 pint).

7 After filling, fit the lid complete with the rubber ring and half-screw or half-snap the fastener.

Testing the seal

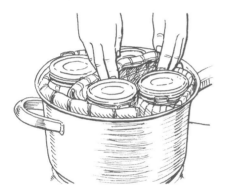

Padding with cloths

8 Place the jars in the pan, and separate them with cloths, so that they will not knock against one another. Then fill the pan with water up to the shoulders of the jars.

9 Slowly heat the water up to 165°F, over a period of between ½ and 1½ hours, depending on the size/texture of the fruit.

10 Hold the temperature for about 20 minutes, remove the jars, screw or snap the lids fully home, and put them on a wooden surface until they are cold.

11 When the jars are completely cold, test the seals by easing back the screw lids or snap fasteners and seeing if the vacuum between the lid/rubber/jar holds. If all is well, screw or snap the lids home, wipe with a clean cloth and store in a cool, dark, frost-free place.

PRESERVING FRUIT BY THE OVEN METHOD

Alternatively, you can preserve fruit by the oven method. Pack the jars with fruit and top up with syrup/water to within 1 in. of the brim. Set the lids in place but do not seal them. With the oven turned up to a temperature of about 250°F, place the jars on a metal tray and put them in the middle of the oven.

After about 30–120 minutes when the fruit has started to bubble – the time depends on the consistency of the fruit (for example, tomatoes take over an hour while plums are done in ten minutes) – remove the jars, tighten the lids and put them on one side to cool down.

MAKING CHUTNEY

The wonderful thing about chutney is that you can, as long as you always use more or less the same base ingredients – salt, sugar and vinegar – design a recipe to suit your own likes and needs. For example, my recipe for apple chutney (see below) has its beginnings in the fact that I love apples and onions but I dislike sultanas. Therefore, when the time came to make chutney, I simply took a traditional recipe and replaced the sultanas with more onions.

The other good thing is that you can change the texture of the chutney to taste. If it is too thick, you can add more vinegar; if it is too thin, you can reduce it by boiling off the water; if it is too dark, you can use white sugar; and so on. Chutney-making gives you a chance to be creative.

HOW TO MAKE APPLE CHUTNEY

This recipes makes about 10–11 lb of apple chutney, but you could increase or decrease the amount, and/or you could change the recipe to marrow, rhubarb or tomato chutney, for example. You will need:

- 6 lb apples.
- 2 lb onions.
- 1 oz salt.
- 1 oz ginger.
- Garlic to taste.
- 1 teaspoon ground mixed spice
- 1 teaspoon cayenne pepper.
- 3 lb brown sugar.
- 2 pints brown malt vinegar.
- As many sterilized glass jars as you can find – jam jars, coffee jars, with or without lids.

- Cellophane or plastic film with wax-paper discs and rubber bands to fit the size and number of your jars.
- Large saucepan or preserving pan – stainless steel, enamel or glass (not iron, copper or brass, and ideally not aluminium).
- Large jam-making funnel.

1 Select and prepare all the ingredients. Wash, peel and core the apples, and peel the onions.

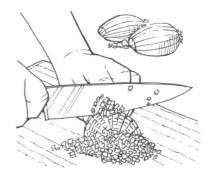

Chopping the onions

2 Chop or mince the ingredients according to taste – for example, I like coarse chutney, while Gill likes hers to be fine-textured, almost like a spread.

3 Put the chopped apples and onions into the pan with either whole spices in a muslin bag, or with sprinkled ground spices. Also add the salt, ginger, garlic and cayenne.

4 Cover with the vinegar.

5 Turn on the heat and gently simmer for 1–4 hours until the contents are soft.

6 Dissolve the sugar in the vinegar and fruit mixture. Slowly bring to the boil, stirring regularly.

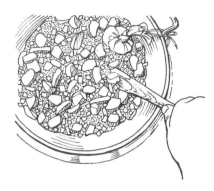

Boiling and stirring

Pouring into jars

Adding the sugar

7 Wash and dry the jars and put them in the oven to warm.

8 Continue simmering and stirring until the whole mixture is the consistency of thin jam – being mindful that chutney thickens as it cools.

9 As soon as the chutney is ready, put the jars on a wooden surface or on newspaper and fill the jars to the top using the funnel.

10 As soon as the jars are full, wipe the rims, fit the covers and store them in a cool, dark airy place. Keep the jars for at least two months before opening, and use within a year.

TROUBLESHOOTING

Unlacquered metal This will corrode if exposed to vinegar. Onions, garlic and other ingredients that need long cooking to tenderize them can, initially, be cooked separately in water in a small pan, because vinegar and sugar tend to harden rather than soften them.

Air pockets Apart from sterilizing and sealing the jars, the main big problem with chutney-making involves making sure that the contents are free from air pockets. This is best achieved by filling the jars with a funnel, and by sharply tapping the jars just before fitting the lids.

MAKING JAM

Making jam is a wonderful activity; one moment you have a glut of fresh fruit – plums, blackberries, apples – and the next you have jars of jam. While bottling fruit can be a disappointing activity, with the fruit variously losing its color, taste and texture, jam-making is a much more satisfying process in which the base fruit is fundamentally transformed.

If you have only ever experienced the uninspiring taste of shop-bought jam, rest assured that home-made jam is a completely different product. It has a sparkling color, the texture is firm, the content is assured, and the taste can be spectacular. As well as that, the activity of collecting the fruit, stirring the pan, pouring the jam into the jars, fitting the lids, and finally eating the jam are all great experiences that should not be missed. Someone once wrote a poem about jars of home-made jam being likened to captured remembrances of summer and autumn – walks along a country lane, picnics in the orchard, the feeling of plucking a ripe plum, the taste and texture of a fresh strawberry – whoever it was must have been a jam-maker.

HOW TO MAKE BLACKBERRY AND APPLE JAM

This recipes makes about 12 lb of blackberry and apple jam, but you could either increase or decrease the amount, and/or you could change the recipe to use gooseberries, just blackberries, plums or whatever might take your fancy.

You will need:

- 6 lb blackberries.
- 2 lb apples.
- A large saucepan or preserving pan – made of stainless steel, enamel or glass (not iron, copper or brass, and ideally not aluminium).
- ½ oz of butter or margarine.
- 1 pint of water.
- 8 lb brown sugar.
- As many sterilized glass jars as you need – jam jars, coffee jars, with or without lids.
- Large jam-making funnel.
- Cellophane or plastic film with wax-paper discs and rubber bands to fit the size and number of your jars.

1 Select and prepare all the ingredients – wash and core the apples, and remove the stalks from the blackberries. The fruit should be firm and ripe. Wash everything, and discard decayed or squashy fruit.

2 Smear the inside of the pan with butter or margarine – this reduces the formation of sugar scum.

3 Weigh the fruit and slide it into the pan along with the water.

4 Turn on the heat, bring to the boil and simmer. Continue until the fruit has broken down.

Adding sugar

Filling the jars

5 Add the sugar little by little, stirring all the time.

6 Continue simmering and stirring until the mix bubbles even when stirred.

Testing for pectin

7 To test, spoon a pool of jam onto a cold plate, allow to thicken for 30 seconds, and then push it with your fingernail. If it wrinkles, it is ready to pot.

8 Turn off the heat. Take the jars (all washed, sterilized, dried and warmed), put them on a wooden surface and use the funnel and a ladle or mug to fill them up.

9 The moment each jar is filled, place the waxed circle wax-side-down on the jam and fit the cover. Place the cellophane cover, which has been dampened on the top side, over the lip of the jar, fit the rubber band, and tweak the cellophane to remove any creases.

Sealing jars

10 Finally, label and date each jar, and store in a cool, dark, dry, airy, mouse- and frost-free cupboard.

181

MAKING BEER

At the first sip of your own home-made beer, you will appreciate that it is a completely different product from the factory-made stuff. While most modern factory-made beers are full of some very nasty additives – all manner of chemicals to variously enrich, color and give the brew body, fizz and character – home-made beers are simply made from water and a few traditional ingredients. If you have looked at the recipe below and are wondering about the shop-bought ingredients, my advice to beginners to beer-making is to start by using carefully selected shop-bought ingredients, and then later, when you understand all the subtleties, gradually change over to using your own ingredients. Much the same goes for the barrel versus bottles issue – it is best for beginners to leave the beer in the barrel rather than putting it into bottles, for the simple reason that it is both easier and safer.

If you have doubts about your skills, have worries about the potential dangers of making large quantities of beer at home, or are thinking that it could be a lonely activity, then one look on the internet or in the local press will confirm that beer-making is so popular that there will almost certainly be one or more local groups where you can swap recipes and equipment, taste other people's beer, and generally exchange ideas. If you are patient, and make sure all the tools and containers are spotlessly clean, then you will not go far wrong.

HOW TO MAKE PALE ALE

You will need:
- Small packet pale ale yeast.
- 5-gallon stainless-steel pan.
- 12 pints water
- 3 lb 8 oz light extract syrup.
- 2 lb light dry extract.
- 8 oz crystal malt.
- 1 oz Fuggles hops – Alpha 50 minutes.
- 1 oz Fuggles hops – Alpha 15 minutes.
- 1 oz Fuggles finishing hops – Alpha dry hops.
- Small packet Irish moss.
- Funnel strainer.
- 5-gallon glass carboy complete with airlock.
- Thermometer.
- 5–7-gallon fermenting bucket.
- Siphon complete with tubing.
- 5-gallon plastic barrel.
- ½ cup sugar.

1 Mix the yeast according to the maker's instructions. If you are using liquid yeast, give the packet a sharp smack to set it into action.
2 Clean all the equipment with a mixture of water and household bleach and rinse thoroughly. Run through the above list – just to make sure all is present and correct.
3 Pour the water into the pan and bring it to a boil.

Adding the malt extract

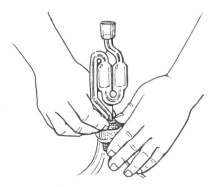

Fitting the airlock

4 Warm the various tins of malt extract. Pour and stir them into the water.

5 Take the Alpha 50 and Alpha 15 hops, read the instructions and then add them to the mix accordingly.

9 Fit the airlock. Put the carboy in a dark room or cupboard for 6–11 days, until the bubbles have more or less come to a halt.

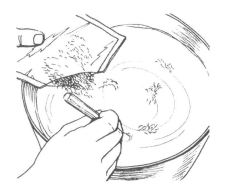

Adding the moss

Siphoning into the barrel

6 Simmer for about 50 minutes, add the moss (the clearing agent) and then the finishing hops.

7 Let the mix – the 'wort' – cool and carefully funnel it into the carboy. Top it up with water to reach the 5-gallon mark.

8 When the temperature is down to about 70°F, add the yeast as described on the packet. Shake the whole mix to aerate.

10 Siphon the brew first into the bucket, so as to leave the sediment behind, and then into the barrel. Add the sugar (dissolved in a cupful of brew) and leave it for 2–6 weeks.

MAKING CIDER

The great thing about cider-making is its simplicity; all you need are apples. Some people do sterilize the juice to kill off natural bacteria and then add another yeast – they claim that it produces a more predictable drink – but it is not the traditional way, and it is not my way. If you collect well-ripened apples, and if you resist the temptation to over-wash them, and if you use a good number of bruised ones, then the cider will effectively make itself.

Traditionally, cider-making was one big party – collecting the apples, lots of singing and romping, drinking lots of juice on the day of pressing, more singing and dancing, celebrating the various tastings, and so on. You can also take joy in the process.

HOW TO MAKE CIDER
You will need:

- Enough windfall apples to make 5 gallon of cider – the quantity needed will depend upon the type of apple and the character of the growing year. Do not worry about bruises and worm-holes – but do throw out anything that is squashy and rotten. Good varieties are Russet, Baldwin, Jonathan and Newton.
- Press/juicer/mincer/liquidizer – anything that will turn apples into pulp.
- Two 5-gallon fermenting buckets, complete with lids.
- Muslin cloth large enough to cover the buckets.
- Room thermometer.
- Siphon complete with tubing.
- 5-gallon plastic barrel complete with airlock.

Chopping the apples

1 Gather your fully ripe windfall apples from clean grass. This is important because, if you have to wash them, there is a danger that you will damage the natural yeasts that are present on the skin. Sort them, discarding mouldy or very brown ones. Roughly chop the remaining apples.

Pressing

2 Crush, press, grind or beat the apples and collect the juice in the buckets – three-quarters fill one bucket and then go on to the next.

Covering with muslin

3 Cover the buckets with the muslin and loosely sit the lids in place.

4 Hold the room temperature at about 60–70°F; check several times a day to ensure that the cider is bubbling but not overflowing.

Siphoning

5 Once the bubbles have flattened out, siphon the cider into the barrel and fit the airlock.

6 Store in a cool dark place and leave for at least three months – six months is better. That said, some makers drink their brew in a few weeks, while others reckon that a good cider should not be made and served in the same year.

BE WARNED

In my experience, cider can be a very heady drink – not at all like lemonade. My advice is to have a little trial sip and then to sit for a moment! As to the practice of bottling cider in recycled wine bottles and adding a little sugar before corking, the big danger is that cheap bottles have a tendency to burst.

CIDER VINEGAR

To make cider vinegar, repeat the procedure as already described, right up to the bubbling stage, and then simply cover the cider with muslin and leave it until it works through and beyond the alcoholic stage, at which point it will turn acid. When this stage is reached, have a trial tasting, and then dilute with filtered rain/spring water to taste. Now, take your favorite herb – something like mint, dill, rosemary or basil – chop it finely and add it to the vinegar.

TROUBLESHOOTING

Too much washing The biggest problem most beginners have is that they hand-pick their apples, and then fiddle and fuss – washing and selecting – to the extent that they kill off all the natural yeasts. Just pick clean windfalls and allow everything in except really rotten apples. The orchard grass must be long and clean with no manure, however.

Health worries If you really are worried about bacteria in the juice, add a Campden tablet to kill it off, add yeast, and continue.

MAKING WINE

The problem most beginners have when it comes to making wine is that they get overly concerned with the jargon and the equipment. You can use French terms, and you can really go to town with thermometers and hydro-meters, and very complex airlocks, and have your own labels printed, and use very special bottles, but you do not have to. You can make beautiful wine simply by following a few basic rules of thumb.

- All the equipment must be clean and sterilized. Some wild yeast does produce good wines, but for the most part they do not.
- Most beginners add too much sugar, to the extent that bottles explode and the wine is a syrupy mess. You must follow the recipes and you are best advised to avoid using bottles until you know what you are doing.
- Make sure that you leave the sugared fruit in a wide-rimmed container – a bucket or tub – not a narrow-necked bottle. The reason for this is that the initial fermentation can be catastrophic with lots of bubbles and fizz.
- Be careful, when you are past the initial fermentation stage, that you avoid using boiling water, because this can kill off the yeast.
- Once the primary fermentation is under way, make sure that the cover is in place, otherwise dust, dirt and flies will get into the wine.
- Make sure when you are siphoning off that you do not siphon up the yeast cells – these are the mess to be found in the bottom of the container.

- Always read the labels when you buy yeast, and always get your supplies from a recommended supplier, where you can get face-to-face advice from an expert.

HOW TO MAKE RHUBARB WINE

You will need:
- 10 lb rhubarb.
- Two 5-gallon fermenting buckets, complete with lid.
- 6 lb sugar.
- Muslin cloth large enough to cover the buckets.
- 2 gallons water
- Large plastic funnel.
- 5-gallon stainless-steel pan.
- 2 lemons, chopped.
- 2 oranges, chopped.
- Large mugful raisins, chopped.
- I oz yeast.
- 2 slices toast.
- Siphon complete with tubing.
- Two fermentation carboys complete with airlocks.
- As many wine bottles as you can find.
- Corks and corking machine.

1 Gather the rhubarb, wash it under running water and cut it into pieces about I in. long.
2 Put the rhubarb into one fermenting bucket along with the sugar, stir it up, cover it with the muslin and the loose lid, and leave it overnight.

Adding the boiled water

Covering with muslin

3 Next day, boil the water and pour it over the sugared rhubarb, cover it as before and let it stand for a week.

4 After seven days, strain it all through the muslin-topped funnel into the stainless-steel pan.

5 Put the mix on to heat and bring it close to the boil; simmer for a few minutes.

9 After about 3–4 days, when the surface is flecked with foam, remove the toast and replace the cover. If it threatens to overflow, pour half into the other bucket.

Adding chopped lemons, oranges and raisins

Siphoning

6 While the mix is still hot but not boiling, add the lemons, oranges and raisins.

7 Mix the yeast according to instructions and stir it into the mix.

8 When the mix is cool, pour it back into the bucket, float the toast on its surface, and cover with the muslin and loose lid.

10 Leave for about 12–15 days, until the dramatic fermentation is complete, and then siphon into one of the carboy jars and fit the airlock.

11 Two weeks later, siphon the wine from one jar into the other and fit the airlock. Repeat this procedure twice more, at 6 and 12 weeks.

12 When the wine is clear, siphon it into the bottles, fit the corks, and leave for at least a year.

MAKING SOAP

HOW TO MAKE SOAP

You will need:

- Camping-type stove.
- I lb tallow (beef fat), or vegetarian alternative.
- Large, stainless-steel pan.
- Muslin cloth.
- ½ cup rainwater.
- I oz lye (sodium hydroxide).
- Petroleum jelly.
- Small, throwaway butter carton, or other mould.

WARNING: L ye (sodium hydroxide) burns; do not get it on your skin. Treat it with extreme caution, as you might neat bleach or acid. Wear old, cover-all clothes, goggles and gloves.

1 On a dry day, set up the camping stove in the garden.
2 Take the tallow, trim off every bit of meat, cut the fat into small pieces, and melt it in the pan. Pour the fat through the muslin and into a bowl to cool.

Cutting the fat

3 Pour the rainwater into a heat-resistant bowl and add the lye to the water. Stir with a wooden spoon until the mixture becomes hot. Then leave it to cool.

Adding the lye

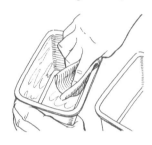

Greasing the moulds

4 While you are waiting for the lye-water mixture to cool, wipe around the inside of the carton or mould with petroleum jelly.

Adding lye to the fat

5 Add the cooled lye to the fat, keep stirring until it becomes like toffee fudge, and then pour it into the carton or mould to set. The lye remains active while the soap is soft, so leave it for 24 hours, in a place well away from children and pets.